QUANTUM REALITY

JIM BAGGOTT

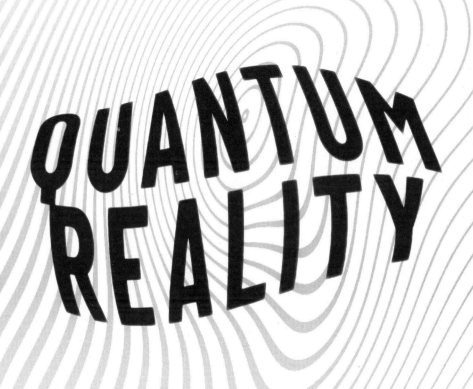

QUANTUM REALITY

OXFORD

UNIVERSITY PRESS

OXFORD
UNIVERSITY PRESS

Great Clarendon Street, Oxford, OX2 6DP,
United Kingdom

Oxford University Press is a department of the University of Oxford.
It furthers the University's objective of excellence in research, scholarship,
and education by publishing worldwide. Oxford is a registered trade mark of
Oxford University Press in the UK and in certain other countries

First Edition published in 2020

Impression: 1

Published in the United States of America by Oxford University Press
198 Madison Avenue, New York, NY 10016, United States of America

British Library Cataloguing in Publication Data
Data available

Library of Congress Control Number: 9780198830160

ISBN 978–0–19–883015–3

Printed and bound in Great Britain by
Clays Ltd, Elcograf S.p.A.

To Ian Mills
Who taught me quite a lot about quantum mechanics

CONTENTS

About the Author   ix

Preamble   xi

Prologue: *Why Didn't Somebody Tell Me About*
All This Before?   1

PART I THE RULES OF THE GAME

1 The Complete Guide to Quantum Mechanics
   (Abridged)   7
   Everything You've Ever Wanted to Know, and a Few
   Things You Didn't

2 Just What is This Thing Called 'Reality', Anyway?   33
   The Philosopher and the Scientist: Metaphysical
   Preconceptions and Empirical Data

3 Sailing on the Sea of Representation   55
   How Scientific Theories Work (and Sometimes Don't)

4 When Einstein Came Down to Breakfast   81
   Because You Can't Write a Book About Quantum
   Mechanics without a Chapter on the Bohr–Einstein Debate

PART II PLAYING THE GAME

5 Quantum Mechanics is Complete So Just Shut Up
   and Calculate   105
   The View from Scylla: The Legacy of Copenhagen, Relational
   Quantum Mechanics, and the Role of Information

6 Quantum Mechanics is Complete But We Need to
   Reinterpret What it Says   127
   Revisiting Quantum Probability: Reasonable Axioms,
   Consistent Histories, and QBism

7 Quantum Mechanics is Incomplete So We Need to
   Add Some Things   151
   Statistical Interpretations Based on Local and Crypto
   Non-local Hidden Variables

8 Quantum Mechanics is Incomplete So We Need to
   Add Some Other Things   175
   Pilot Waves, Quantum Potentials, and Physical Collapse
   Mechanisms

9 Quantum Mechanics is Incomplete Because
   We Need to Include My Mind (or Should That be
   Your Mind?)   199
   Von Neumann's Ego, Wigner's Friend, the Participatory
   Universe, and the Quantum Ghost in the Machine

10 Quantum Mechanics is Incomplete Because…
   Okay, I Give Up   221
   The View from Charybdis: Everett, Many Worlds, and
   the Multiverse

   Epilogue: *I've Got a Very Bad Feeling about This*   249

   Appendix: Realist Propositions and the Axioms of
      Quantum Mechanics   255
   Acknowledgements   257
   List of Figure Acknowledgements   259
   Endnotes   261
   Bibliography   283
   Index   293

ABOUT THE AUTHOR

Jim Baggott is an award-winning science writer. A former academic scientist, he now works as an independent business consultant but maintains a broad interest in science, philosophy, and history and continues to write on these subjects in his spare time. His previous books have been widely acclaimed and include the following:

The Quantum Cookbook: *Mathematical Recipes for the Foundations of Quantum Mechanics*

Quantum Space: *Loop Quantum Gravity and the Search for the Structure of Space, Time, and the Universe*

Mass: *The Quest to Understand Matter from Greek Atoms to Quantum Fields*

Origins: *The Scientific Story of Creation*

Farewell to Reality: *How Fairy-tale Physics Betrays the Search for Scientific Truth*

Higgs: *The Invention and Discovery of the 'God Particle'*

The Quantum Story: *A History in 40 Moments*

Atomic: *The First War of Physics and the Secret History of the Atom Bomb 1939–49*, short-listed for the Duke of Westminster Medal for Military Literature

A Beginner's Guide to Reality

Beyond Measure: *Modern Physics, Philosophy, and the Meaning of Quantum Theory*

Perfect Symmetry: *The Accidental Discovery of Buckminsterfullerene*

The Meaning of Quantum Theory: *A Guide for Students of Chemistry and Physics*

PREAMBLE

I know why you're here.

You know that quantum mechanics is an extraordinarily successful scientific theory, on which much of our modern, tech-obsessed lifestyles depend, from smartphones to streaming to satellites. You also know that it is completely mad. Its discovery forced open the window on all those comfortable notions we had gathered about physical reality from our naïve interpretation of Isaac Newton's laws of motion, and unceremoniously shoved them out. Although quantum mechanics quite obviously *works*, it appears to leave us chasing ghosts and phantoms, particles that are waves and waves that are particles, cats that are at once both alive and dead, lots of seemingly spooky goings-on, and a desperate desire to lie down quietly in a darkened room.

But, hold on. If we're prepared to be a little more specific about what we mean when we talk about 'reality' and a little more circumspect about how we think a scientific theory might *represent* such a reality, then *all the mystery goes away*.

I'm not kidding. I have a bit of a reputation as the kind of guy you might find in the kitchen at parties; the kind who spoils all the fun, bursting the bubbles of excitable mystery and urban myth (what Americans sometimes call 'woo') with a cold scepticism and a calculating rationality. Spock, not Kirk (or McCoy). One commentator recently called me 'depressingly sane'.* This is a badge I'm happy to wear with pride. There are

---

* This was theoretical physicist Sabine Hossenfelder, referencing my book *Farewell to Reality: How Fairy-tale Physics Betrays the Search for Scientific Truth*, in a tweet dated 11 March 2018.

many popular books you can buy about the weirdness and the 'woo' of quantum mechanics. This isn't one of them.

And in any case that's not why you're here.

But—let's be absolutely clear—a book that says, 'Honestly, there is no mystery' would not only be a bit dull and uninteresting (no matter how well it was written), it would also be completely untrue. For sure we can rid ourselves of all the mystery in quantum mechanics but only by abandoning any hope of deepening our understanding of nature. We must become content to use the quantum representation simply as a way to perform calculations and make predictions, and we must resist the temptation to ask: *But how does nature actually do that?* And there lies the rub: for what is the purpose of a scientific theory if not to aid our understanding of the physical world?

Let's be under no illusions. *The choice we face is a philosophical one.* There is absolutely nothing scientifically wrong with a depressingly sane interpretation of quantum mechanics in which there is no mystery. If we choose instead to pull on the loose thread we are inevitably obliged to take the quantum representation at face value, and interpret its concepts rather more literally. Surprise, surprise. The fabric unravels to give us all those things about the quantum world that we find utterly baffling, and we're right back where we started.

My purpose in this book is (hopefully) not to spoil your fun, but to try to explain what it is about quantum mechanics that forces us to confront this kind of choice, and why this is entirely philosophical in nature. Making different choices leads to different interpretations or even modifications of the quantum representation and its concepts, in what I call (with acknowledgements to George R. R. Martin) *the game of theories.*

Part I opens with a brief summary of everything you might need to know about quantum mechanics, which should hopefully set you up for what follows. I will then tell you about the rules of the game, based on a pragmatic but perfectly reasonable understanding of what we mean by 'reality', and the kinds of things we can hope to learn from a scientific representation of this. Part I concludes with Albert Einstein's great debate with Niels Bohr in the late 1920s and early 1930s, and the emergence of the anti-realist Copenhagen interpretation, which admirably sets the scene.

We will then go on in Part II to look at various attempts to play the game, from the legacy of Copenhagen, through relational quantum mechanics, to interpretations based on quantum 'information'. We will look at attempts to redefine quantum probability, by reformulating the axioms of quantum mechanics, introducing the notion of consistent histories, and quantum Bayesianism. We then turn our attention to realist interpretations based on the idea that quantum mechanics is a statistical theory. These include hidden variable theories of local (Bell's inequality), and 'crypto' non-local (Leggett's inequality) varieties.

Experimental evidence gathered over the past forty years or so comes down pretty firmly against local and crypto non-local hidden variables. So we turn to interpretations based on non-local hidden variables (such as so-called 'pilot wave' theories) or we try to fix problems associated with the 'collapse of the wavefunction' by introducing new physical mechanisms, including a possible role for human consciousness. We conclude with the notion that the wavefunction is real but doesn't collapse, which leads to many worlds and the multiverse.

If you will indulge me, through all of this I will make use of a no doubt overfanciful analogy or metaphor.[1] This is based on the notion that the game of theories involves navigating the 'Ship of Science' on the perilous 'Sea of Representation'. Yes, I've obviously read too many fantasy novels.

We sail the ship back and forth between two shores. These are the deceptively welcoming, soft, sandy beaches of Metaphysical Reality and the broken, rocky, and often inhospitable shores of Empirical Reality. The former are shaped by our abstract imaginings, free-wheeling creativity, personal values and prejudices, and a variety of sometimes pretty mundane things we're obliged to accept without proof in order to do any kind of science at all. These become translated into one or more *metaphysical preconceptions*, which summarize how we think or even come to believe reality should be. These are beliefs that, by their nature, are not supported by empirical evidence. So, if you prefer you could think of these preconceptions as intuitions or even articles of faith, echoing one of my favourite Einstein quotes: 'I have no better expression than the term "religious" for this trust in the rational character of reality and in its being accessible, to some extent, to human reason.'[2]

Within the sea I have charted two grave dangers. The rock shoal of Scylla lies close to the shores of Empirical Reality. It is a rather empty instrumentalism, perfectly empirically adequate but devoid of any real physical insight and understanding. Charybdis lies close to the beaches of Metaphysical Reality. It is a whirlpool of wild, unconstrained metaphysical nonsense. The challenge to theorists is to discover safe passage across the Sea of Representation. In *Quantum Reality* I want to explain why this has proven so darn difficult, and why I have a very bad feeling about it.

So, welcome. You're here because you want some answers. Please take a seat and make yourself comfortable, and I'll go and put the kettle on.

PROLOGUE

Why Didn't Somebody Tell Me About All This Before?

My first encounter with quantum mechanics occurred in my very first term as an undergraduate, studying for a bachelor's degree in chemistry in a rather damp and gloomy Manchester, England, in the autumn of 1975.

Looking back, it's no real surprise that all the students in my class (me included) were utterly baffled by what we were taught. Until that moment, we had all been blissfully unaware that there was anything more to be learned about the physical world beyond the smooth continuity and merciless certainties of Newton's clockwork mechanics.

Our understanding of atoms was limited to the 'planetary model' associated with the names of physicists Niels Bohr and Ernest Rutherford. If we had thought about it at all (and I can tell you that we really hadn't), then we would have supposed that the classical theories we use to describe planets orbiting the Sun could simply be extended to describe little balls of electrically charged matter orbiting the central nucleus of an atom. Yes, the forces are different, but surely the results would be much the same.

But now we were told that the physics of atoms and molecules is governed by a very different set of laws, with which even chemists must come to terms. Nothing had prepared us for this. In our first lecture we chomped our way through Max Planck's discovery of the quantum, Einstein's 'light-quantum' hypothesis, Bohr's quantum theory of the atom, Louis de Broglie's wave-particle duality,* Erwin Schrödinger's wave mechanics, and Werner Heisenberg's uncertainty principle.

I thought my head was going to explode.

Mechanics is that part of physics concerned with the how and why of stuff that *moves*, governed by one or more mathematical *equations of motion*. In hindsight, our problems were compounded by the fact that the evolution of our understanding of classical mechanics had stopped with the school textbook version of Newton. We were not being trained to be physicists, and so missing from our education was the elaborate reformulations of classical mechanics, first by Joseph-Louis Lagrange in the eighteenth century, and then by William Rowan Hamilton in the nineteenth. These reformulations weren't simply about recasting Newton's laws in terms of different quantities (such as energy, instead of Newton's mechanical force). Hamilton in particular greatly elaborated and expanded the classical structure and the result, called *Hamiltonian mechanics*, extended the number of situations to which the theory could be applied.

We were therefore confronted not only with this extraordinary thing called the quantum *wavefunction*, but also with the challenge of writing down something called the 'Hamiltonian' for a specific physical system or situation, such as the orbit of an electron in an atom or the vibrations of a chemical bond holding

---

* More than forty years later, I can still hear my lecturer pointing out as an aside that de Broglie is pronounced 'de Broy'.

two atoms together, without really understanding where either of these things had come from.*

But, make no mistake, I was completely hooked. I filled my notebooks with equations that looked…well, they looked *beautiful*. I still didn't really understand what any of it meant, but I learned how to use quantum mechanics as best I could and set aside any concerns. I went on to complete a doctorate at Oxford University and a couple of years of postdoctoral research at Oxford and at Stanford University in California, before returning to England to take up a lectureship in chemistry at the University of Reading. Although I was never blessed with any great ability in mathematics, I learned a great deal more about quantum mechanics from Ian Mills, professor of chemical spectroscopy in my department, and I take some pride in a couple of research papers I published on the quantum theory of high-energy molecular vibrations.

Then, in 1987, whilst working for a couple of months as a guest researcher at the University of Wisconsin-Madison, I happened upon an article that sent me into a tailspin. This was written by N. David Mermin.[1] It told of something called the Einstein–Podolsky–Rosen 'thought experiment', which dates back to 1935, and some laboratory experiments to probe the nature of quantum reality that had been conducted by Alain Aspect and his colleagues in 1982.

I felt embarrassed. I had come to this really rather late. *Why didn't somebody tell me about all this before?*

I had allowed my (modest) ability in the *use* of quantum mechanics to fool me into thinking that I had actually understood

---

* I've written a technical book, suitable for readers with a background in physical science and some capability in mathematics, called *The Quantum Cookbook: Mathematical Recipes for the Foundations of Quantum Mechanics*. This was published by Oxford University Press in 2020 and I consider it a 'companion' to this volume. It is the book that I would have found really helpful when I was 18.

it. Mermin's article demonstrated that I really didn't, and marked the beginning of a 30-year personal journey. I'm now the proud owner of several shelves overflowing with books on quantum mechanics, science history, and philosophy, and a laptop filled with downloaded articles. I've written a few books of my own, the first published in 1992.

I can happily attest to the fact that, like charismatic physicist Richard Feynman, I still don't understand quantum mechanics.[2] But I think I now understand *why*.

PART I
THE RULES OF THE GAME

1

THE COMPLETE GUIDE TO QUANTUM MECHANICS (ABRIDGED)

Everything You've Ever Wanted to Know,
and a Few Things You Didn't

Here's what I've learned over the past forty years or so.

Nature is lumpy, not smooth and continuous

We now know that all matter is composed of atoms. And each atom is in turn made up of light, negatively charged electrons 'orbiting' a nucleus consisting of heavy, positively charged protons (two up quarks and a down quark), and electrically neutral neutrons (one up quark and two down quarks).[*] Atoms are discrete. We can say that they are 'localized'. Atoms are 'here' or 'there'. In itself this is not particularly revelatory.

But despite what a few ancient Greek philosophers had argued, two and a half thousand years before, towards the end of the nineteenth century atoms were really rather controversial. After

---

[*] I've put 'orbiting' in inverted commas because the electron doesn't orbit the nucleus in the same way that the Earth orbits the Sun. In fact, it does something a lot more interesting, as we'll soon see.

all, why believe in the existence of atoms when you can never hope to see them or gain any kind of evidence for them?

In fact, it was a determination to refute the existence of atoms that led Max Planck to study the properties and behaviour of so-called 'black-body' radiation trapped inside cylindrical vessels made from platinum and porcelain.* When such a vessel is heated, its interior glows like a furnace. As the temperature rises, the light radiation released inside glows red, orange-yellow, bright yellow, and ultimately brilliant white. Planck was interested in finding a theory to describe the variations in the pattern and intensities of different frequencies (or wavelengths, or colours) of the radiation as the temperature is raised.

What Planck found in an 'act of desperation' in 1900 turned him into a committed atomist, but it took a few more years for the real significance of his discovery to sink in. Planck had concluded that the radiation inside the cavity is absorbed and emitted by its walls as though it is composed of discrete bits which he called *quanta*. This is summarized in an equation now known as the Planck–Einstein relation:

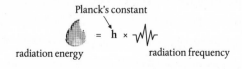

This might not seem very profound. But think about it. Radiation frequency is smoothly and continuously variable—there are no sudden jumps or breaks in the rainbow spectrum of colours: instead they blend seamlessly from one colour to another. If

---

* 'Black-body' doesn't refer in any way to the colour of the walls of the cavity, but rather to the way they absorb and emit the radiation trapped inside. In theory, a 'black' body absorbs and emits radiation 'perfectly', meaning that the radiation doesn't depend on what the walls are made of.

energy = radiation frequency, then this would imply that energy too must be smooth and continuous. But this is not what Planck had found. For any given frequency Planck's constant (represented by the letter h) represents the smallest amount of energy that can be absorbed or emitted by an object. Energy isn't taken up or given out smoothly and continuously by the object, but rather in discrete bits determined by h. Planck's constant is the telltale sign of all things quantum.

Planck initially attributed this behaviour to the atomic nature of the material that formed the walls of the vessels. But it was Einstein who really launched the quantum revolution when in 1905 he rather outrageously suggested that the *radiation itself* is 'quantized' in localized, discrete bits or lumps of energy. This is Einstein's 'light-quantum' hypothesis, and it is the reason why we refer to the above relation today using both Planck and Einstein's names. He was right, of course. We now know these lumps of light-energy as *photons*.

So it's not only matter that comes in lumps, but also radiation. Put more and more energy into an electron inside an atom and it will 'orbit' the nucleus at greater and greater average distances until it is ripped out of the atom completely. But you can't increase this energy smoothly and continuously. The electron will absorb energy only at very discrete intervals organized in an *atomic spectrum* (see Figure 1).

These intervals form a ladder with rungs in a distinctive pattern. It was Niels Bohr in 1913 who figured out that this pattern is characterized by one or more *quantum numbers* and, unlike a real ladder, the quantum rungs get closer and closer together the higher in energy you go. Pump just the right amount of energy into an electron in an atom, sufficient to climb from one rung to the next, and the electron's orbit appears to change *discontinuously*, in a 'quantum jump'.

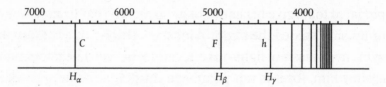

Figure 1 This picture shows a series of lines in the atomic spectrum of hydrogen, which consists of a single proton orbited by a single electron. Energy increases from left to right, and the spectrum shows that energy is not absorbed or emitted continuously, but only in discrete amounts. This spectrum appears in a 1910 textbook *Lærebog i Physik*, by Christian Christiansen, who taught Niels Bohr at the University of Copenhagen. Wavelengths are recorded in ångstroms (tenths of a nanometre or billionth of a metre) along the top, with the characteristic spectral lines H_α (656.3 nanometres—red), H_β (486.1 nanometres—blue) and H_γ (434.0 nanometres—violet) clearly marked.

So far as we know, there is nothing in reality that isn't quantized, possibly including space and time.

Waves are particles and particles are waves

I have to admit that French physicist Louis de Broglie is a bit of a hero of mine. He seems to have had little impact on science after winning the Nobel Prize in 1929, but his contribution six years earlier was more than enough to leave a lasting mark on human history.

Einstein's light-quantum hypothesis was greeted with considerable scepticism at the time. When he was recommended for membership in the prestigious Prussian Academy of Sciences in 1913, its leading members—Planck among them—acknowledged his remarkable contributions to physics, which by this time included his special theory of relativity (the general theory would follow a few years later). In accepting his nomination, they were prepared to forgive his lapses of judgement: 'That he may some-times have missed the target in his speculations, as, for example, in his hypothesis of light-quanta, cannot be really held too much against him, for it is not possible to introduce really new ideas even in the most exact sciences without sometimes taking a risk.'[1]

But in his short paper, Einstein had suggested that it might be possible to gain evidence for the quantum nature of light by studying the *photoelectric effect*. Shine light on metal surfaces of a certain frequency and intensity and electrons will get kicked out. Now, the energy of a classical wave is related to its amplitude—the height of its peaks and depths of its troughs—think of the difference between gently rolling surf and a tsunami. This energy is reflected in the *intensity* of the wave or, if you prefer, its *brightness*. If, as everybody believed, light is described purely in terms of waves, then increasing the light intensity increases the energy and should therefore smoothly increase the number and ener-gies of the electrons ejected from the surface.

But this is not what was observed in early experiments. The Planck–Einstein relation suggests that it is the light frequency—not the intensity—that is all important. Light of the wrong fre-quency, no matter how intense, just won't cut it. Only light-quanta (photons) with sufficient energy will knock the electrons from the surface. Increasing the intensity of the light simply increases the number (but not the energies) of the ejected electrons.

At the time this kind of behaviour was very counterintuitive, but it was nevertheless shown to be correct in further experi-ments performed about ten years later, and led to the award of the Nobel Prize in Physics to Einstein in 1921.

This was a great achievement, but it also posed a big problem. There was an already well-established body of evidence in favour of a wave theory of light. Push light of a single colour through a narrow aperture or slit, cut with dimensions of the order of the wavelength of the light, and it will squeeze through, bend around at the edges and spread out beyond. It 'diffracts'. A photographic plate exposed a short distance away will reveal a diffuse band, rather than a narrow line with the same dimensions as the slit.*

Cut two slits side by side and we get *interference*—evidenced by alternating bright and dark bands called interference fringes. As the waves from both slits spread out and run into each other, where wave peak meets wave peak we get a bigger peak (which we call constructive interference), and where peak meets trough we get cancellation (destructive interference)—see Figure 2a. Constructive interference gives rise to the bright fringes. Destructive interference gives rise to the dark fringes. This kind of behaviour is not limited to light—such wave interference is fairly easily demonstrated using water waves, Figure 2b.

But waves are inherently *delocalized*: they are here *and* there. Einstein's light-quantum hypothesis didn't negate all the evidence for the delocalized wave-like properties of light. What he was suggesting is that a complete description somehow needs to take account of its localized particle-like properties, too. He had some ideas about how this might be done, and we'll come to consider these later on in this book.

Okay, so light exhibits some peculiar behaviours, but matter must surely be different. It's fairly straightforward to show that material particles such as electrons behave pretty much how we would expect. For example, we observe distinct tracks in a device called a cloud chamber—see Figure 3a.† This picture shows a

---

* However, look closely and you'll see that the edges of this band show a distinct *diffraction pattern* of alternating light and dark 'fringes'.
† The cloud chamber was invented by Charles Wilson. It works like this: an energetic, electrically charged particle passes through a chamber filled with

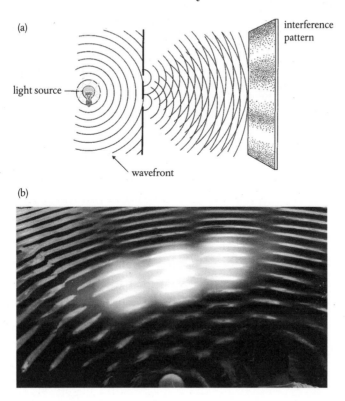

Figure 2 (a) When passed through two narrow, closely spaced slits, light of a single wavelength produces a pattern of alternating light and dark fringes. These can be readily explained in terms of a wave theory of light in which overlapping waves interfere constructively (giving rise to a bright fringe) and destructively (dark fringe). (b) Such interference is not restricted to light, and can be fairly easily demonstrated with water waves.[2]

bright track left by a positively charged alpha particle (the nucleus of a helium atom, consisting of two protons and two neutrons), and a series of fainter tracks left by negatively charged electrons, their curved motions caused by the application of a magnetic field.

vapour. As it passes, it dislodges electrons from atoms in the vapour, leaving charged ions in its wake. Water droplets condense around the ions, revealing the particle trajectory.

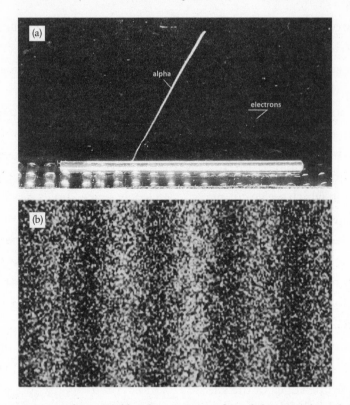

Figure 3 (a) Tracks caused by the motions of an alpha particle and electrons ejected from a rod of radioactive thorium inside a cloud chamber. (b) A two-slit interference pattern generated using electrons.

The simplest explanation for such tracks is that they trace the paths or *trajectories* of individual particles as they pass through the chamber.

And here we come to de Broglie's historically profound insight. Why force a distinction? If light waves can also be particles (photons, though this name hadn't yet been coined), *could particles like electrons also be waves?* The idea seems completely absurd and, indeed, some physicists dismissed it as 'la Comédie Française'. We're so used to thinking of elementary particles like electrons as small, localized bits of charged matter that to imagine them any other way requires some considerable mental effort.

Readers familiar with televisions in the days before plasma and LCD screens may recall that these consisted of one or more electron or 'cathode ray' guns, each of which would produce a beam of electrons. The beams were then accelerated and modulated to produce broadcast images on a phosphorescent screen.

So, imagine we pass a narrow beam of electrons through a plate in which we've cut two small, closely spaced holes or slits.

Our instinct might be to imagine that in a two-slit experiment, the electrons in the beam will follow paths through either one slit or the other, like machine-gun bullets, producing two bright lines on the screen marking where the electrons have passed through. We would expect each line to be brightest in the centre, showing where most of the electrons have passed straight through the corresponding slit unimpeded, becoming a little more diffuse as we move away, signalling electrons that have caught the edges of the slit and scattered on their way through. But these experiments have been done, and this is not what we see. Instead of two bright lines characteristic of particles following straight paths through the slits, we get a two-slit interference pattern—Figure 3b.

Electrons can also be waves.

De Broglie's idea was just that—an idea. He was able to develop a direct mathematical relationship between a wave-like quantity—wavelength—and a particle-like quantity—linear momentum*—such that

---

* In classical mechanics, linear momentum derives from the uniform motion of an object travelling in a straight line, calculated as the object's mass × velocity. However, in quantum mechanics, the calculation of linear momentum is rather different, as we'll see very soon.

$$\text{wavelength} \quad \overset{\text{Planck's constant}}{\underset{\text{linear momentum}}{\mathcal{N} = \frac{h}{\text{\tiny }}}}$$

But this was not a fully fledged wave theory of matter. That challenge fell to Erwin Schrödinger, whose formulation—first published early in 1926 and called wave mechanics—is still taught to science students today.

Everything we think we know about a quantum system is supposed to be summarized in its wavefunction

Schrödinger's theory is really the classical theory of waves in which we make use of the de Broglie relation to substitute wavelength for linear momentum. This requires a bit of mathematical sleight of hand and some assumptions that prove to be unjustified. Although Schrödinger published a much more obscure derivation, this is what it really boils down to. The result is *Schrödinger's wave equation*.

It's helpful to stop and think about this for a minute. The classical wave equation features a *wavefunction*, which you can think of as describing a familiar sine wave, oscillating smoothly and continuously between peak and trough. The wave equation then describes the motion of this wave in space and time. Into this, we've now injected Planck's constant and linear momentum, a very particle-like property. If we adopt the classical expression for momentum as mass × velocity, you can see that this is now a wave equation that features something that has a mass, and this gets incorporated into the solutions of the equation—the wavefunctions.

How can a wave have mass? This is just one mind-bending consequence of wave–particle duality. And we're just getting started.

What's quite fascinating about all this is that, right from the very beginning, physicists were scratching their heads about Schrödinger's wavefunction. It's pretty obvious how the wavefunction should be interpreted in classical wave theory but, aside from now featuring particle-like properties such as mass and momentum, in Schrödinger's wave mechanics the wavefunction had taken on an altogether different significance.

In classical mechanics, there are no real issues with the way we interpret the concepts represented in the theory. We think we know what mass is. We know what velocity and acceleration are. These are things we *observe* directly—by simple observation we can tell the difference between something moving slowly and something moving fast. When we put our foot down and go from nought to sixty in some incredibly short time, or when we loop the loop on a rollercoaster, we *feel* the acceleration. We can calculate linear momentum and we know what this means. These things, called physical 'observables', sit right on the surface of the classical equations of motion. We don't have to dig any deeper for some kind of hidden meaning for them. It's obvious what they are and how they should be interpreted.

But now look at what Schrödinger's wave mechanics asks us to do. You want to know the linear momentum of an electron moving freely through a vacuum? Then you need to solve the wave equation and identify the relevant wavefunction, determine the rate of change of this wavefunction in space, and multiply the result by minus the square root of minus 1 times Planck's constant divided by 2π.* This procedure returns the linear momentum

---

* The square root of minus 1 is an 'imaginary number', usually written as i. This might seem obscure, but it crops up all the time in mathematics and physics. All you need to remember is that $i^2 = -1$.

multiplied by the wavefunction, from which we can then deduce the momentum.

In Schrödinger's wave mechanics (and, more generally, in quantum mechanics), we calculate observables such as momentum and energy by performing specific mathematical *operations* on the relevant wavefunction. Such manipulations are then summarized collectively as *operators* for the observables. The operators are mathematical recipes, which we can think of as 'keys' which unlock the wavefunction (depicted below as a box), releasing the observable before closing again. The logic is as follows:

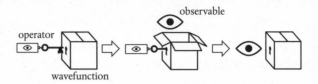

The description given in the paragraph above summarizes the mathematical operator (the key) for linear momentum in quantum mechanics. There's one further small step. I won't give the details here, but it is quite straightforward to deduce something called the *expectation value* of the operator, which is a kind of average value. It has the helpful property that

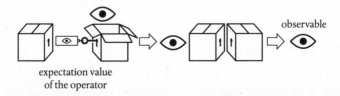

When two identical boxes face off against each other as 'mirror images', as shown above, and if everything is being done properly, then these combine together to produce the result 1. This leaves us with just the observable, so the expectation value provides a useful recipe for calculating the values of observables such as momentum and energy.

Whoa. You don't need to be a rocket scientist to realize that something has fundamentally changed. It's as though nature has chosen to hide its secrets in the quantum wavefunction, hence the locked box pictogram. To discover the value of an observable, we need to open the box with the right key (the operator). Opening the box with one kind of key gives us one kind of observable, such as momentum. A different observable will require a different key.

We never had to do anything like this in classical mechanics. The observables were always right there, in front of us, staring us in the face.

No, seriously, electrons really do behave like waves

At this point I want to bring you back to the electron interference pattern shown in Figure 3b. We might shrug our shoulders at this, and acknowledge the wave nature of electrons without thinking too deeply about what this might mean. But let's push the experiment a stage further. Let's wind down the intensity of the electron beam so that, on average, only *one* electron passes through the slits at a time. What then?

What we see is at first quite comforting. Each electron passing through the slits registers as a single bright dot on the phosphorescent screen, telling us that 'an electron struck here'. This is perfectly consistent with our prejudices about electrons as particles, as it seems they pass—one by one—through one or other of the slits and hit the screen in a seemingly random pattern—see Figure 4a.

But wait. The pattern isn't random. As more and more electrons pass through the slits we cross a threshold. We begin to see individual dots group together, overlap, and merge. Eventually we get a two-slit interference pattern of alternating bright and dark fringes, Figure 4e.

We can quickly discover that if we close one or the other slit or try to discover which slit each individual electron passes through then we will lose the interference pattern. We just get behaviour characteristic of particles following straight-line paths. If we try to look to see how we get wave behaviour, we get particle

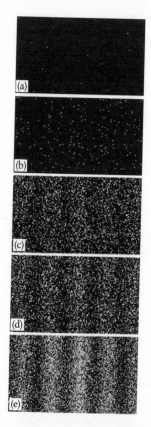

Figure 4 We can observe electrons as they pass, one at a time, through a two-slit apparatus by recording where they strike a piece of photographic film. Photographs (a) to (e) show the resulting images when, respectively, 10, 100, 3,000, 20,000, and 70,000 electrons have been detected.

behaviour. If we don't look to see how we get wave behaviour, we get wave behaviour. Left to itself, it seems that the behaviour of each electron must somehow depend on the existence of the slit through which it *does not pass*, which is decidedly odd.

Alternatively, we conclude that the wave nature of the electron is an *intrinsic* behaviour. *Each individual electron behaves as a wave*, described by a wavefunction, passing through both slits simultaneously and interfering with itself before striking the screen.

So, how are we supposed to know precisely *where* the next electron will appear?

The wavefunction gives us only probabilities: in quantum mechanics we can only know what might happen, not what will happen

A wave alternates between positive amplitude, largest at a peak, and negative amplitude, largest at a trough. We calculate the intensity of the wave as the *square* of its amplitude, which is always a positive number. So, in two-slit interference described purely in terms of waves we imagine a resulting wave which, when squared, produces a pattern which alternates between regions of high intensity (bright fringes) and zero intensity (dark fringes), as shown in Figure 5a.

But, by its very nature, this pattern of intensity is spread across the screen. It is *distributed* through space, or delocalized. And yet we know that in the experiment with electrons, as illustrated in Figure 4, we see electrons detected one at a time, as single bright spots, in only *one* location on the screen. Each electron hitting the screen is localized.

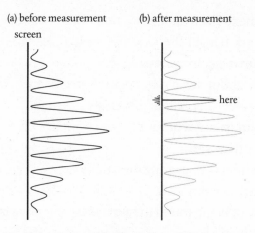

(a) before measurement   (b) after measurement

screen

here

Figure 5 (a) Before measurement, the square of the electron wavefunction predicts a distribution of probabilities for where the electron might be found, spread across the screen. (b) After measurement, the electron is recorded to be found in one, and only one, location on the screen.

How does this work?

Schrödinger had wanted to interpret the wavefunction literally, as the theoretical representation of a 'matter wave'. He argued that atoms are simply the diffraction patterns of electron waves captured and wrapped around atomic nuclei. But to make sense of one-electron interference we must reach for an alternative interpretation suggested later in 1926 by Max Born.

Born reasoned that in quantum mechanics the square of the wavefunction is a measure not of the intensity of the electron wave, but of the *probability* of 'finding' its associated electron.[*] The alternating peaks and troughs of the electron wave translate into a pattern of quantum probabilities—in this location (which will become a bright fringe) there's a higher probability of

---

[*] To be clear, because the wavefunction might contain i, the square root of -1, we multiply it by its *complex conjugate*, in which i is replaced by $-i$ (since $-i \times i = -i^2 = +1$), so we always get a positive result. This is called *modulus square* of the wavefunction. It is actually the two locked boxes facing off against each other.

finding the next electron, and in this other location (which will become a dark fringe) there's a very low or zero probability of finding the next electron.

Just think about what's happening here. Before an electron strikes the screen, it has a probability of being found 'here', 'there', and 'most anywhere' where the square of the wavefunction is bigger than zero.

Does this mean that an individual electron can be in more than one place at a time? No, not really. It is true to say that it has a probability of being found in more than one place at a time and there is definitely a sense in which we think of the electron wavefunction as delocalized or distributed. But if by 'individual electron' we're referring to an electron as a particle, then there is a sense in which this *doesn't exist* as such until the wavefunction interacts with the screen, at which point it appears 'here', in only one place, as shown in Figure 5b.

That this might be a bit of a problem was recognized in the late 1920s/early 1930s by John von Neumann. If 'measurement' is just another kind of quantum process or transition, then von Neumann argued that this suggests a need for a 'measurement operator', such that

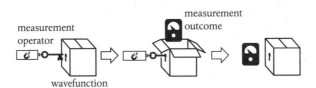

The measurement outcome is then just the expectation value of the measurement operator.

Just like the distributed interference pattern shown in Figure 5a, the wavefunction in question may consist of different measurement possibilities, such as the pointer of the gauge above pointing to the left or to the right. Von Neumann realized that there is nothing in the mathematical structure of quantum mechanics that explains how we get from many *possible* outcomes to just one *actual* outcome. So, to ensure that the structure is mathematically robust and consistent, he had no choice but to *postulate* a discontinuous transition or jump which gets us from the possible to the actual. This postulate is generally known today as the 'collapse of the wavefunction'. It is absolutely central to the ongoing debate about how quantum theory is to be interpreted.

Quantum probability is not like classical probability

One more thing. That there's a 50% probability that a tossed coin will land 'heads' simply means that it has two sides and we have no way of knowing (or easily predicting) which way up it will land. This is a classical probability born of ignorance. We can be confident that the coin continues to have two sides—heads and tails—as it spins through the air, but we're ignorant of the exact details of its motion so we can't predict with certainty which side will land face up.

Quantum probability is thought to be very different. When we toss a quantum coin* we might actually be quite knowledgeable about most of the details of its motion, but we can't assume that 'heads' and 'tails' *exists* before the coin has landed, and we look.

Einstein deplored this seeming element of pure chance in quantum mechanics. He famously declared that 'God does not play dice'.[3]

---

* We'll see how this can be done in practice in later chapters.

For a specific physical system or situation, there is no such thing as the 'right' wavefunction

Physics is a so-called 'hard' or 'exact' science. I take this to mean that its principal theoretical descriptions are based on rigorous mathematics, not on words or phrases that can often be ambiguous and misleading. But mathematics is still a language, and although we might marvel at its incredible fertility and 'unreasonable effectiveness',[4] if not applied with sufficient care it is still all too capable of ambiguity and misdirection.

Centuries of very highly successful, mathematically based physics have led us to the belief that this is all about getting *the right answer*. Nature behaves a certain way. We do *this*, and *that* happens. Every time. If the mathematics doesn't predict *that* with certainty every time we do *this*, then we're inclined to accept that the mathematical description isn't adequate, and we need a better theory.

In quantum mechanics, we're confronted with a few things that might seem counterintuitive. But this is still a mathematically based theory. Sure, we've swopped the old classical observables such as momentum and energy for mathematical operators which we use to unlock their quantum equivalents from the box we call the wavefunction. But—to take one example—the frequencies (and hence the energies) of the lines in an atomic spectrum are incredibly precise—just look back at Figure 1. If quantum mechanics is to predict what these should be, then surely this must mean discovering the precise expression for the wavefunction of the electron involved?

And it is here that we trip over another of quantum mechanics' dirty little secrets. There is really no such thing as the 'right' wavefunction. All we need is a function that is a valid solution of the wave equation. Isn't this enough to define the 'right' one? No, not really. Whilst there are some mathematical rules we need to respect, we can take any number of different solutions and combine them in what's known as a *superposition*. The result is also a perfectly acceptable solution of the wave equation.

I want to illustrate this with an example from the quantum theory of the hydrogen atom, consisting of a nucleus formed by a single proton, 'orbited' by a single electron. In fact, this was the problem that Schrödinger addressed in his 1926 paper with such spectacular success. The wavefunctions of lowest energy form spherical patterns around the central nucleus. But there are wavefunctions of modest energy that are shaped like dumbbells. There are three of these.

All three of these solutions of the wave equation are characterized by a set of quantum numbers. Two of these are the same for each of the dumbbell-shaped functions, but the third differs from one to the other, taking values of −1, 0, and +1, as shown in Figure 6a. For now it doesn't really matter what these quantum numbers represent. Here's the thing. Whilst these are the 'natural' solutions of the wave equation, they're not the most helpful when we come to think about combining atoms in three-dimensional space to form molecules which, after all, is what *chemistry* is all about.

It's much easier to deal with wavefunctions defined in three spatial dimensions, using Cartesian x, y, and z coordinates. This is okay for the function characterized by the quantum number 0, as we can simply define this to lie along the z coordinate. But what of the others? Well, this turns out to be relatively easy. To get a wavefunction directed along the y coordinate we form a superposition in which we *add* the functions corresponding to +1 and −1, as shown in Figure 6b. To get a wavefunction directed along the x coordinate we form a superposition in which we *subtract* the function corresponding to +1 from the function corresponding to −1. Because we're combining functions that are

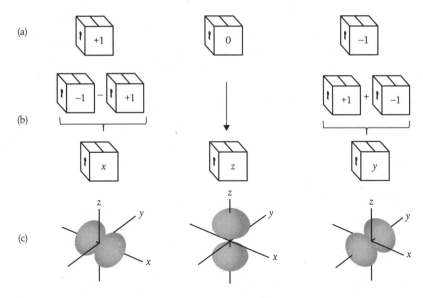

Figure 6 The 'natural' solutions of the Schrödinger wave equation for the hydrogen atom include a set of wavefunctions characterized by quantum numbers with values +1, 0, and −1. But it's often more helpful to combine these in the way shown here, which produces three wavefunctions directed along the three spatial dimensions characterized by Cartesian coordinates x, y, and z. So, which are the 'right' wavefunctions?

known to be solutions of the wave equation, and provided we follow the rules, we can be confident that the superpositions represent valid solutions, too. The resulting functions are shown mapped along the three coordinates in Figure 6c.

But which then are the 'right' wavefunctions? Students learn fairly quickly that there really isn't a straightforward answer to this question. The 'right' wavefunction obviously depends on what kind of system we're dealing with, but we're free to choose the form that's most appropriate for the specific problem we're trying to solve.

Delocalized waves can be combined together in ways that localized particles simply can't, and we can take full advantage of this in quantum mechanics.

Heisenberg's uncertainty principle is about what we can know. It is not about what we can only hope to measure

Werner Heisenberg hated Schrödinger's wave mechanics. Not because the mathematics was dodgy, but because Schrödinger insisted that the wavefunction be taken literally as the physical description of the electron as a matter wave. The trouble is that waves flow smoothly and continuously, and there is no room in this picture for sudden quantum jumps of the kind needed to interpret the transitions that form an atomic spectrum. Schrödinger simply doubled down and denied that the jumps happened at all, proclaiming: 'If all this damned quantum jumping were really here to stay, I should be sorry I ever got involved with quantum theory.'[5]

In 1927, Heisenberg realized that the essential discontinuity—the 'jumpiness'—at the heart of quantum mechanics implies a fundamental limit on what we can discover about the values of pairs of physical observables, such as position and linear momentum. At first, he believed that this limit arises because of an inevitable 'clumsiness' involved in making measurements on delicate quantum systems with our large-scale, laboratory-sized instruments. For example, Heisenberg argued, if we want to determine the precise position of an electron in space we need to locate it by hitting it with photons of such high energy that we must forgo any hope of determining the electron's precise momentum. Any attempt to 'see' where the electron is (or, at least, was) will just knock it for six, preventing us from seeing to where and how fast it was going.

But Bohr disagreed, and the two argued, bitterly. Bohr insisted that the uncertainty principle has nothing to do with the clumsiness or otherwise of our measurements. Instead it implies a fundamental limit on what we can *know* about a quantum system.

Perhaps the simplest way of explaining Bohr's point relies on the essential duality of waves and particles in quantum mechanics. Think about how we might measure the wavelength of a wave. We could infer the wavelength by counting the numbers of peaks and troughs in a certain fixed region of space. Each wave cycle consists of one peak and one trough, and the wavelength is the distance from the start to the finish of the cycle. So we sum the numbers of peaks and troughs, and divide by two. This tells us the number of cycles in our spatial region. The wavelength is then the length of this region divided by the number of cycles.

Obviously, we will struggle to make any kind of precise measurement if our sample region is *shorter* than the wavelength. We quickly realize that we can increase the precision by making the region large enough to include lots and lots of cycles. Now, from the de Broglie relation, a precise wavelength gives us a precise measure of linear momentum for the associated particle. But, of course, we've deliberately made our sample region large, so we've lost any hope of measuring a precise position for the associated particle. It could be anywhere in there.

The opposite is also true. It's possible to add together a large number of waves in a superposition, each with a different wavelength, such that the wave has a single peak located at a very precise position. Such a superposition is called a 'wavepacket'. This gives us a fix on the precise position, but now we've lost any hope of measuring a precise momentum, because our wavepacket consists of a broad range of wavelengths, implying a broad range of momenta.

Bohr's view prevailed, and we now write Heisenberg's uncertainty relation as[*]

---

[*] The sign ≥ means 'is greater than or equal to'. This works in the same way as an equals sign. For example, we could divide both sides by the uncertainty in momentum to give: uncertainty in position ≥ $h/(4\pi$ × uncertainty in momentum).

$$\text{uncertainty in position} \times \text{uncertainty in linear momentum} \geq \frac{h}{4\pi} \quad \text{Planck's constant}$$

Note that nowhere does this say that measurements of position and momentum are somehow mutually exclusive. We can in principle measure the position with absolute precision (zero uncertainty), but then the momentum would be completely undetermined: it would have infinite uncertainty. There's nothing preventing us from determining both position and momentum with more modest precision within the bounds of the uncertainty principle.

The principle is not limited to position and momentum. It applies to other pairs of observables. Perhaps the best known relates to energy and time, which we will meet again in Chapter 4. Now, there's a caveat. There are arguments that the energy–time uncertainty relation actually doesn't exist except as a variation of that for position and momentum. Early attempts to derive the energy–time relation from first principles proved rather unsatisfactory. To my knowledge, the most widely accepted derivation, published in 1945 by Leonid Mandelstam and Igor Tamm, clearly specifies the interpretation of 'time' in the relation as a time *interval*.

I think this is enough for now. I want you to be clear that what I've described so far is based on the 'authorized' or 'official' version of quantum mechanics taught to science students all around the world. We'll see in Part II that as the search for meaning has unfolded in the past 90 years or so, some physicists and

---

This shows straight away that if the uncertainty in momentum is zero, then the uncertainty in position would be infinite.

philosophers have happily challenged this authority, and we shouldn't assume that the version taught today will still be taught in another 90 years' time.

Eager readers will also note that I've deliberately held back some of the more infamous examples of quantum weirdness—such as Schrödinger's cat and the Einstein–Podolsky–Rosen experiment. Please be patient: we will come to these in Chapter 4. I first want to give you some context in which to think about them.

To summarize, we've seen that experimental discoveries in the first decades of the twentieth century led to the realization that physical reality is inherently lumpy. In the classical mechanics of everyday life, we can safely ignore Planck's constant and the lumpiness it implies, and assume everything is smooth and continuous. But at the level of molecules, atoms, and subatomic particles, from which everything in the visible universe is constructed, Planck's constant comes into its own and we can no longer ignore the duality of waves and particles.

De Broglie opened Pandora's Box in 1923. Schrödinger gave us his wave equation and his wavefunctions a few years later. The all-too-familiar observables of classical mechanics became locked away inside the quantum wavefunction, requiring mathematical operators to liberate them from their prison. Born said that the wavefunctions are utterly inscrutable; they tell us only about quantum probabilities. Heisenberg (and Bohr) explained that the heart of quantum mechanics beats uncertainly. Nature suffers a peculiar arrhythmia.

And then the debates began. What is quantum mechanics telling us about the nature of physical reality? And just what is this thing called reality, anyway?

2

JUST WHAT IS THIS THING CALLED 'REALITY', ANYWAY?

The Philosopher and the Scientist: Metaphysical Preconceptions and Empirical Data

It's a cliché, but as I've grown older I've definitely become more impatient, more of an old curmudgeon. I've grown increasingly exasperated with our species' seeming inability to learn anything at all from the past. This is certainly not an exasperation born of a nostalgia for the 'good old days'. As anyone who grew up in the shadow of the Cold War and the threat of mutual assured destruction will tell you, there is no fondness for those times among those who survived them. Rather, this is an exasperation born from watching helplessly as the seemingly common sense and virtuous victories of decades of peaceful, liberal democracy are unravelled by a new breed of aggressive, opportunistic populists, intent on exploiting the legacy of a decade of economic hardship inflicted in 2008 by the greed of a few bankers. We seem to be at great risk of forgetting everything we've learned from modern humanity's great successes and its horrific past mistakes.

My impatience knows no bounds. Much the same goes for our understanding of 'reality'. I can't begin to count the number of books and research papers I've read, the number of television documentaries I've watched, which purport to tell us how this or that scientific theory describes something new and unusually bizarre about reality, without ever being clear on the kind of reality that is supposedly being described.

Hang on. 'Kind of reality'? What on Earth is that supposed to mean?

If you pick up an introductory text on philosophy, there's a good chance that this will have sections on things like epistemology (the study of knowledge and justified belief), metaphysics (the study of the fundamental nature of being and existence, encompassing ontology and cosmology), logical reasoning, the philosophy of mind, moral philosophy, ethics and aesthetics, and the philosophy of science. If you look up 'reality' in the index, you'll find that this is a subject discussed extensively under the heading of *metaphysics*.

So, to a philosopher, reality is metaphysics (meaning, literally, 'beyond physics'). And yet, to a physicist, reality is something described by theories that are unquestionably *scientific*, such as quantum mechanics. What's going on?

Now, there have been some well-known and very well-respected scientists who have publicly disparaged philosophy, 'which at its best seems to me a pleasing gloss on the history and discoveries of science', writes Nobel laureate Steven Weinberg.[1] This kind of negativity is born from a judgement that philosophy appears to offer little or no guide to the inception, development, and evolution of a scientific theory, or even what it really means when we talk about the 'scientific method'. 'Let's not put the cart before the horse,' says Stanford University theorist Leonard Susskind, 'Science is the horse that pulls the cart of philosophy.'[2] 'Of course, philosophy is the field that hasn't progressed in two thousand years, whereas science has,' says astrophysicist Lawrence Krauss, in a talk about his book *A Universe from Nothing*, an example of rather poor quality philosophy masquerading as science.[3]

Rather like the fractious members of the People's Front of Judea in Monty Python's *Life of Brian*, such scientists demand to know: 'What has philosophy ever done for us?'

I confess I don't really understand this kind of argument, or the attitudes that lie beneath it. So let's be absolutely clear. Philosophy is *not* science. Philosophy does make progress, but this is not the same kind of progress that we tend to associate with science. It's likely true that we can't use philosophy to tell us *how* to develop a scientific theory, though I believe it holds some useful lessons, as we'll see in the next chapter. This is, after all, what *science* is supposed to be for. But, as I hope to be able to show in what follows, when we look closely we find that it's not actually possible to do science of any kind without metaphysics, interpreted broadly as *the assumption of things we can't prove*. And the moment we accept this is the moment we open the door to philosophy.

As we've seen in the opening chapter, quantum mechanics forces us to confront some uncomfortable truths about what we can and cannot hope to fathom about the nature of reality. I firmly believe that if we want to understand how to *interpret* what quantum mechanics is telling us, then appreciation of a few lessons from philosophy is absolutely essential.[4]

So, let's get back to reality. We'll start by trying to discover what—if any—difference exists between the realities of the philosopher and the scientist.

Fans of the 1999 movie *The Matrix* will recall the scene in which Morpheus instructs Neo: 'How do you define real? If you're talking about what you can feel, what you can smell, what you can taste and see, then real is simply electrical signals interpreted by your brain.'[5] Spend a few minutes pondering on this, and you should have no real difficulty in accepting its basic truth. You rely on your body's sensory apparatus to deliver a complex set of sense impressions (converted into electrical signals) to your brain. As a result of some processes we do not yet fully understand, these impressions are synthesized in your conscious mind to deliver a set of perceptions and experiences (what some philosophers refer to as 'qualia') which combine to form a representation of the world around you. The result is what you call your reality. It is very specifically—and uniquely—yours.

You take this at face value because you have no real choice. If you went around questioning everything you perceive, then you wouldn't get much done. Is this rose really red? Just what is 'red' anyway? Such questions concern your conscious experience, but what if you're not around to experience things? If a tree falls in a forest, and there's nobody around to hear, does it still make a sound? Is the Moon still there if you're not looking at it, or thinking about it? And there goes another missed deadline.

You might want to argue that, after all, you are a highly intelligent life form, the result of about four billion years of evolution by natural selection on planet Earth, punctuated by at least five mass extinctions. Does it make any sense for *Homo sapiens* to have evolved a way of perceiving a reality that is in some way fundamentally different from, or inconsistent with, how it *really is*?

But then in another moment of quiet reflection you will realize that there is no evolutionary law you can point to that would guarantee this. One of the factors that contributes to the survival of a species in the hectic scramble we call life is that certain genetic mutations bring with them survival advantages that are then selected for. With any luck, you live long enough to procreate, hopefully passing these advantages to a new generation. All we can be sure of is that we've evolved a finely tuned mental *representation* of those aspects of reality necessary to ensure our survival. There is no evolutionary selection pressure to develop a mind to represent reality as it really is.

Unsure? Ponder the evidence from *synesthesia*, a condition in which those who experience it* report perceptions that have become 'mixed up', with stimulation of one sense triggering involuntary responses from one or more other senses. One fairly common form is known as grapheme-colour synesthesia, in which letters and numbers are perceived to be coloured: 'Wednesday is indigo blue'.[6] It's easy to dismiss this as incorrect wiring or 'cross-talk' in the brain, giving rise to an incorrect representation of reality, but with a prevalence estimated to be about four per cent of the population, those affected do not always see it this way. Who's to say whose perceptions are the 'right' ones?

And, before you ask, it's no good trying to corroborate what you perceive by sharing your experiences with a friend. Unless they're a synesthete (and you're not), your friend will doubtless confirm that they perceive precisely the same things that you perceive: 'Yes, that rose is red.' But, even if we presume that your friend has a mind that works in much the same way as yours (and he or she is not a philosophical zombie), all this tells us is that both your minds have developed in a similar fashion. You learned about the colour red as a small child, perhaps from pictures in a book that your parents would point to whilst saying 'red' out loud. You can be pretty sure that your friend's knowledge and understanding of colour is derived from a set of very similar experiences. All this tells us is that your minds have undergone much the same conditioning, producing what philosopher John Searle refers to as the 'background'.

---

* I'm reluctant to write 'those who suffer from it' as some synesthetes regard themselves as gifted.

This problem with reality has been recognized by philosophers since the ancient Greeks. In *The Republic*, which was written nearly two and a half thousand years ago, Plato devised an allegory which we can re-purpose to explore the situation a little more deeply. This is Plato's famous allegory of the cave.

Deep in the cave is a number of prisoners, chained to a wall. They have lived their entire lives in the cave, and have no experience of a world outside. They're not even aware that they are prisoners.

It is dark, but the prisoners are nevertheless aware of men and women passing continually along the wall in front of them, carrying all sorts of vessels, and statues and figures of animals. Some are talking among themselves. But, in truth, there is a fire constantly burning at the back of the cave, filling it with a dim light. The fire is out of sight and the prisoners are completely unaware of it. The men and women that the prisoners can see against the far wall are, in fact, the *shadows* cast by real people passing in front of the fire. The reality that the prisoners experience is made up of the crude *appearances* of things—people and objects—which they have mistaken for the things themselves.

Plato's allegory was intended to illustrate his three-tier theory of knowledge. The shadows represent common belief or popular opinion (*doxa*) based merely on appearances. The objects themselves represent a deeper form of understanding derived, for example, from science (*episteme*, from which we get 'epistemology'). The topmost tier is *noesis* (or 'nous'), knowledge that goes beyond the superficial facts of the objects and concerns their form and nature.[7] But the allegory serves our purpose here as an illustration of the fact that, as human beings, we rely on our senses to deliver a representation of reality that we have learned to take at face value, if not for granted. In our everyday lives, it's simply a waste of time to question everything we experience.

We should nevertheless acknowledge a simple truth. Our reality is made up of shadows, of things-as-they-appear, and we have no real way of knowing to what extent the representation shaped by our perceptions reflects reality as it really is, a reality of things-in-themselves.

So how then can we be sure that a reality of things-in-themselves even exists? Well, we can't, but let's fast-forward a couple of thousand years to 1781, and the great philosopher Immanuel Kant. In the *Critique of Pure Reason*, Kant distinguished between what he called *noumena*, the metaphysical objects of reality or things-in-themselves that we can conceive only in our minds, and the empirical *phenomena*, the shadows or the things-as-they-appear in our perception and experience.[*]

Don't take this to mean that noumena are merely figments of a fertile imagination. I can imagine all kinds of entertaining things—such as Gandalf, unicorns, or Westeros—but these obviously don't connect with phenomena; they don't manifest themselves as things we can directly perceive, except in works of fiction.

More practically, we can point to lots of ways in which what we call 'electrons' manifest themselves in our empirical reality. These are electrons-as-they-appear. But an electron-in-itself without any kind of interaction through which it can make itself manifest exists, kind of by definition, only in our imaginations.

Kant claimed that it makes no sense to deny the existence of the things-in-themselves, as there must be some things that cause appearances in the form of sensory perceptions: there can

---

[*] I should point out that identifying the noumenon with Kant's 'thing-in-itself' (*Ding an sich*), is still debated today among philosophers and somewhat controversial. For a good discussion, see Sebastian Gardner, *Kant and the Critique of Pure Reason* (Routledge, London, 1999), pp. 200–1.

be no appearances without anything that appears.* Just because we can't perceive reality as it really is doesn't mean that it has ceased to exist. The great science-fiction writer Philip K. Dick was surely paraphrasing Kant when he observed: 'Reality is that which, when you stop believing in it, doesn't go away.'[8]

But we have to accept a trade-off. Whilst we might happily conclude that the things-in-themselves must exist, we must grumpily accept that we can in principle gain no knowledge of these. When judged in terms of Plato's three tiers, Kant denies that noesis is possible—we can never have knowledge of the form and nature of the things-in-themselves.

Now, in the last chapter we encountered another big disconnect, of a very different kind but no less profound. This is the disconnect between the quantum world of molecular, atomic, and subnuclear dimensions and the classical world of everyday experience. It is the disconnect created by the locked box of the wavefunction, measurement operators, and the collapse of the wavefunction. What we will discover is that our anxiety over the relationship between reality and perception carries over to that between reality and *measurement*. We will find that we can no longer assume that what we measure necessarily reflects reality as it really is, and that there is also a difference between things-in-themselves and things-as-they-are-measured.

---

* Kant's *Critique* is tough going, and it's clear that his take on phenomena is not as simple and straightforward as I'm suggesting here. Phenomena are not 'derived' from noumena in the same way that shadows are cast by the objects in Plato's cave. Our senses don't think, and the human mind can intuit nothing without sensory perception. To gain knowledge we need to bring the two together. Phenomena are then 'conceptualized appearances' determined by the way the mind works, including our intuitions of space and time. Kant argued that such appearances are nevertheless objective, as demanded for the 'synthetic unity of thought'.

The contemporary physicist and philosopher Bernard d'Espagnat called it 'veiled reality', and commented that 'we must conclude that physical realism is an "ideal" from which we remain distant. Indeed, a comparison with conditions that ruled in the past suggests that we are a great deal more distant from it than our predecessors thought they were a century ago.'[9]

So, this is why philosophers consider any kind of speculation, any conception, discussion, dissection, or thesis on the nature of a reality of things-in-themselves, to be metaphysics.*

It's probably about here that more pragmatic readers might be starting to lose patience. Philosophers are known for their tendency to argue, obfuscate, and confuse, to see problems where they don't exist and make mountains out of molehills. This stuff about noumena and phenomena is all very well, but scientists don't want to waste their precious time nitpicking over the meanings of words. 'Physics', the philosopher(!) Karl Popper once said in an interview, 'is that!', as he slammed a book down hard on the table in front of him.[10]

Science is surely different. It proceeds through the painstaking gathering of hard, reproducible, and verifiable *facts*. Scientists develop theories that can accommodate these facts and explain the patterns that they form in terms of some underlying laws of nature. These theories make predictions that can be tested by reference to new observations or experiments which generate new facts. Sometimes the new facts don't fit, so the theory is either tweaked in some way or thrown out and replaced by a new theory. This is how science makes progress in ways that (at least according to Krauss) philosophy doesn't.

---

* At least since Kant and, arguably, since Plato. Of course, this doesn't stop some contemporary philosophers (and, as we will see later in this book, some theoretical physicists) arguing that pure metaphysics can still somehow generate objective knowledge.

Except that it's not quite as simple as this, as we'll see in the next chapter.

Okay, so we can never be sure that the reality that we perceive or measure reflects or represents the things-in-themselves, but this quite obviously doesn't prevent us from making observations, doing experiments, and developing and testing scientific theories. We can still determine that if we do *this* then *that* will happen. We can still establish hard facts about the shadows—the projections of whatever we think reality might be into our world of perception and measurement—and we can compare these with similar facts that have been derived by others. If these facts agree, then surely we have learned something about the nature of reality?

And this is indeed the bargain we make. The philosophers tell us that reality-in-itself is metaphysics. Although scientists don't often openly acknowledge it up front, the reality that they study is inherently an *empirical reality* deduced from their studies of the shadows. It is an empirical reality of observation, experiment, measurement, and perception; an empirical reality of things-as-they-appear and of things-as-they-are-measured. As Heisenberg once explained: 'we have to remember that what we observe is not nature in itself but nature exposed to our method of questioning.'[11]

We're not quite done yet. Scientists work best within a framework based on laws or rules. They need parameters. Even though we might not be able to gain knowledge of a metaphysical reality-in-itself, we can follow Kant (and Philip K. Dick) and *assume* that such a reality must exist. There can be no appearances without something that appears. What's more, it must surely exist *objectively* and *independently* of our ability to perceive or measure it. We would expect that the shadows would continue to be cast on the wall whether or not there were any prisoners in the cave to observe them.

We might also agree that, whatever reality is, it does seem to be rational and predictable, at least within limits. It appears to be logically consistent. The shadows that we perceive and measure are surely not completely independent of the things-in-themselves that cause them; otherwise, anything goes and science of any kind would be impossible. Even though we can never have knowledge of the things-in-themselves, as Kant argued, we can *assume* that the properties and behaviour of the shadows they cast are somehow determined by the things that cast them.

The truth is that empirical reality is a pretty dull place. It is a reality consisting only of numbers, of effects, a reality of doing this and getting that. Imagine a research paper that says: 'We did these things and we got these results.' Full stop, end of story. The numbers and the effects are meaningless until we try to *interpret* them. To do this we construct a theory that tries to explain what's going on, with the broad aim of improving our understanding. And I would argue that any kind of scientific theorizing is simply impossible without first assuming the independent existence and the rational, logical consistency of the reality that lies beneath all the bald empirical experiences.

What is the justification for these assumptions? At great risk of repetition, here's that Einstein quote once again: 'I have no better expression than the term "religious" for this trust in the rational character of reality and in its being accessible, to some extent, to human reason.'[12]

Scientists generally don't think too long and hard or look too closely at these assumptions. Many regard them as intuitively obvious. But accepting these assumptions means buying into a specific *philosophical position*. This has a name—it's called 'scientific realism'. Curiously, those who don't think about it at all can't avoid making the same kind of commitment, knowingly or not, typically to an alternative but related position called 'naïve realism', a term which I've always thought to be rather pejorative. Like Monsieur Jordain in Molière's *The Bourgeois Gentleman*, who discovers that he has actually been speaking prose all his life, so might some scientists discover that, at least in part, throughout their career in science they've also adopted a specific philosophical position, and so they've been doing philosophy.

Knowingly or not, most scientists are realists, although many don't trouble themselves with grand questions about the nature of reality, which they'll happily leave to philosophers. Although Einstein's philosophy evolved through his early life, as far as quantum mechanics is concerned he was a realist,* as was Schrödinger (and Popper). Many years ago I trained as an experimental scientist and I can tell you it's really difficult to work in a laboratory and maintain your sanity without some belief in the reality of the things you're experimenting with.

So, let's put our realist stake in the ground with a proposition (philosophers like propositions):†

Realist Proposition #1: *The Moon is still there when nobody looks at it (or thinks about it).*

This will actually save us quite a lot of time. In essence, it says that reality (whatever this might be) really does exist and, what's more, it exists independently of our ability to make empirical observations and measurements on it. It continues to exist whether or not anybody is thinking about it. It is *objective*, not subjective. It doesn't depend on me (or you) for its existence.

---

* Einstein read Kant's *Critique of Pure Reason* at the age of 13, and for a time Kant was his favourite philosopher.
† I'll be making frequent references to these propositions throughout the rest of the book. To save you the trouble of looking back here to remind you what they are, I've collected them all together in a handy appendix at the back of this book.

It's important to acknowledge that, no matter how reasonable Proposition #1 might seem to you, it is nevertheless an *assumption*. What's more, it's an assumption that you will never be able to prove. Accepting this means admitting metaphysics into the very foundations of science.

Now we have to do something about the fact that quite a lot of science (and pretty much all of quantum mechanics) deals with things we can't observe directly, but for which we gain empirical evidence that is *indirect*. These are things like photons, electrons, and quarks. These produce effects in our empirical reality of experience—such as interference patterns, lines in a cloud chamber, and 'jets' of hadrons observed at the Large Hadron Collider at CERN, which can be traced back to the behaviour of quarks and gluons, the force particles that hold quarks together inside protons and neutrons. We choose to interpret these effects in terms of 'invisible' quantum entities, and we ascribe to them physical properties such as mass, electrical charge, spin, flavour, and colour. We then explain the observable, empirical behaviour in terms of the properties we have ascribed to these invisible entities.

Let's agree that, based on the above discussion, these properties tell us only about the things-as-they-appear or the things-as-they-are-measured, and we have no way of acquiring knowledge of photons-in-themselves or electrons-in-themselves, and so on. But this doesn't prevent us from assuming that the things-in-themselves really do exist, independently of any instrument required to make measurements on them. For example, we see patterns of behaviour that we explain in terms of electrical charge. But electrical charge is merely the empirical manifestation of whatever property 'in reality' gives rise to such effects. This could very well be electrical charge as we understand it, but the truth is we have no way of knowing.

I propose to focus here on the reality or otherwise of the entities themselves, and set aside consideration of their properties for the next chapter. One of my favourite arguments for 'entity realism' comes from the philosopher Ian Hacking. In an early passage of his 1983 book *Representing and Intervening*, Hacking describes the details of a series of experiments designed to discover if it is technically possible to reveal the fractional electric charges characteristic of 'free' quarks (the answer, alas, is no). The experiments involved studying the flow of electric charge across the surfaces of balls of superconducting niobium:[13]

> Now how does one alter the charge on the niobium ball? 'Well, at that stage,' said my friend, 'we spray it with positrons to increase the charge or with electrons to decrease the charge.' From that day forth I've been a scientific realist. *So far as I'm concerned, if you can spray them then they are real.*

Whilst Hacking is a realist about invisible entities, this doesn't mean that he accepts that scientific theories about such entities are necessarily 'true'. We'll go on to consider the scientific representation of properties and behaviours in the next chapter, but as far as entity realism is concerned I discover that I can find no better words:

Realist Proposition #2: *If you can spray them then they are real.*

This should save us quite a bit of time, too.

You might be tempted to think that we're now faced with a simple choice between the metaphysical reality of the philosopher and the empirical reality of the scientist. Some scientists and philosophers have indeed argued that this is a straightforward, black-and-white choice between metaphysics and empiricism. An outright rejection of metaphysics led arch-empiricist Ernst

Mach to conclude that there could be no such things as atoms. In the 1920s and 1930s, a group called the Vienna Circle, led by philosophers Moritz Schlick, Rudolph Carnap, and Otto Neurath (and others), concluded that all metaphysics should be rejected as meaningless. They attempted to establish a scientifically based philosophy in which experience is the only valid source of knowledge. This is *logical positivism*, a kind of 'seeing-is-believing' brand of philosophy, which was directly influenced by Mach.[14]

But the fact is that Mach got atoms hopelessly wrong and the logical positivist programme failed.[15] It is actually impossible to eliminate all metaphysics from science and, incidentally, this kind of realization is one way in which it is possible to see progress in philosophy. This is another reason why scientists inadvertently end up doing at least some philosophy, whether they are aware of it or not.

The problems arise just as soon as we try to go beyond the empirical evidence—the numbers, the effects, the doing this and getting that. Going beyond the evidence means opening the door to metaphysics. For the simple reason that it is impossible to engage in any meaningful dialogue about 'why?' and 'how?' without first developing some notions about the reality that we assume to lie beneath the empirical data. Scientists choosing this path have no choice but to indulge their inner metaphysicians.

This brings me rather neatly to the analogy I mentioned in the Preamble. To understand how science is done we need to recognize and respect the very distinct ways of thinking about reality adopted by scientists and philosophers, and the fundamentally important relationship that exists between them. I've tried to achieve this by conceiving of them as shores, separated by a sea, as illustrated in Figure 7.

Giving flight to fancy, I think of the shores of Metaphysical Reality as idyllic. This is a warm, sunny, and welcoming place; of

Figure 7 A metaphor for scientific theorizing: sailing the Ship of Science across the Sea of Representation, between the shores of Metaphysical and Empirical Reality.

soft, sandy beaches and lush tropical vegetation. It is a place of abstract imagination and absolute perfection. Here you will find conceptions for how reality might be, could be, or should be. These are conceptions born from the personal values and prejudices of the individual scientists who visit, based on everything they have been taught and have come to understand about the nature of reality, and many things they can only guess at.

This is the place where scientists can be human—more Kirk than Spock. Here they can express their emotions, set free their desires, vent their frustrations, and just be themselves. Here it is possible to believe without evidence, to accept something as true without proof, to have faith. Some may look to find God here (in which case this place is a kind of heaven). Those scientists who prefer to deny the existence of God may instead look here to find nothing at all, because (for them, at least) conjuring something from nothing by physical mechanism alone represents the ultimate triumph of scientific (but actually just another kind of metaphysical) reasoning over religious superstition.

Such metaphysics might involve grand visions of reality, its cause, nature, and ultimate fate. But, on average, it is more likely to be concerned with some of reality's rather more mundane aspects, the things-in-themselves and the 'nuts and bolts' required to assemble reality and hold it together.

You might be doubtful that scientists really need this place. Why should they want to come here? Surely they have everything they need in Empirical Reality? So let me try to illustrate why scientists find themselves needing to make frequent visits.

As we saw in Chapter 1, driven by otherwise inexplicable evidence, physicists of the early twentieth century were obliged to torture quantum mechanics from the classical mechanics that preceded it. As a result, our understanding of physical concepts such as linear momentum was profoundly and irreversibly changed. *But some classical concepts were retained.*

For one thing, the classical mechanics we inherited from Newton requires an *absolute* space and time. This is a problem because, if it existed, an absolute space would form a curious kind of container, presumably of infinite dimensions, within which some sort of mysterious cosmic metronome marks absolute time. It implies a vantage point from which it would be possible to look down on the entire material universe, a 'God's-eye view' of all creation. The arch-empiricist Mach rejected this notion too, and Einstein arguably (and finally) dispensed with it in 1915 in his general theory of relativity, in which space and time emerge as dynamic variables.

Absolute space and time or, more correctly, absolute *spacetime*, is a metaphysical construction. And yet here it is still, dragged largely unchanged from classical mechanics to form the backdrop to a quantum mechanics, against which we imagine quantum events play out. Knowingly or not, anyone doing quantum mechanics is accepting the notion that reality consists of an absolute spacetime. They accept this without the benefit of any empirical evidence, its truth (or at least its acceptance) established without the need for proof.

Scattered along these shores you will also find the abstract, metaphysical concepts needed to lubricate the mathematical machinery of science. These are things such as the perfect circle and the perfect sphere; the infinitesimally small point; the infinitesimal angle; the infinite straight line; the natural limit; and many other things besides. These concepts do not exist in Empirical Reality, and yet without them we would struggle to do mathematically based science of any kind.[16]

For scientists who spend even a short amount of time here, all this gets bundled up into a set of *metaphysical preconceptions*. These summarize how scientists think about reality and the kinds of things they believe it should possess. They include preconceptions that might be based on abstract notions such as beauty, symmetry, elegance, or simplicity. They give rise to convictions that theoretical explanations should be 'natural'. Or that reality must be deterministic, with effect always following inevitably from cause. Or that there are some unassailable 'natural laws' (for example, of conservation) that must always be respected. Or that spacetime must be absolute.[17]

Across the sea we find the shores of Empirical Reality. Now, this is the reality of our everyday experience, so I should apologize up front if you find the way I paint it to be intolerably bleak. Aside from anything else, I want to contrast Empirical Reality with the sandy, sylvan, alluring loveliness of Metaphysical Reality, and I'm mindful of Humphrey Davy's homily about science being a harsh mistress, delivered to the impressionable young Michael Faraday, who on 1 March 1813 was newly appointed to a position at London's Royal Institution.*

So, the shores of Empirical Reality are rocky and frangible, tormented by high winds and lashing waves. This is the inhospitable home of hard, if not rather brutal, empirical facts, of numbers and effects, of doing this and getting that. It is of course here where we discover how nature *actually* appears. This is where (supposedly according to Benjamin Franklin) gangs of brutal facts have been known to murder beautiful theories or, according to Thomas Huxley, we witness the 'great tragedy of science— the slaying of a beautiful hypothesis by an ugly fact'.[18] This is where scientists must set their emotions and their prejudices aside and practise a cold and calculating rationality—definitely more Spock than Kirk.

But don't think for a minute that there's no room for metaphysics, even here. We are obliged to accept that no observation

* See http://www.rigb.org/blog/2013/march/faraday-appointment

or experiment is possible unless we have at least some idea of what we're looking at, or experimenting on. Of course this doesn't mean we can never discover new things, just that it is impossible to make an observation or a measurement without the context of a supporting theory of some kind, though that theory might change or be replaced entirely as a result.

French physicist and philosopher Pierre Duhem once suggested that we go into a laboratory and ask a scientist performing some basic experiments on electrical conductivity to explain what he is doing:[19]

> Is he going to answer: 'I am studying the oscillations of the piece of iron carrying this mirror?' No, he will tell you that he is measuring the electrical resistance of a coil. If you are astonished, and ask him what meaning these words have, and what relation they have to the phenomena he has perceived and which you at the same time perceived, he will reply that your question would require some long explanations, and he will recommend that you take a course in electricity.

Empirical facts are never theory-neutral; they are never free of contamination from some theory or other. As we construct layer upon layer of theoretical understanding of phenomena, the concepts of our theories become absorbed into the language we use to describe the phenomena themselves. Facts and theory become hopelessly entangled: what we observe in Empirical Reality depends to a certain extent on how we look at it.

No wonder the logical positivists failed.

The shores of Metaphysical and Empirical Reality are separated by a sea. As I explained earlier, if we are to make sense of the empirical facts by developing explanations which deepen our understanding, we must find ways to bring together our preconceptions and the facts. We must create some tension between them. We therefore set sail in the Ship of Science, bringing all our metaphysical preconceptions with us and using our knowledge of the empirical facts to develop a workable *representation* of reality, or what we call a scientific theory. Our purpose in crossing the Sea of Representation is to find a way to accommodate both our preconceptions and the facts in a single, typically (but not exclusively) mathematical structure.

There are many different ways of making the journey across the sea, suggesting that it is indeed possible to develop many different theories, each with a different mix of preconceptions, and each of which provide perfectly adequate accounts of the empirical data. This is known to philosophers of science as the *underdetermination of theory by data*. It is a bit of a disappointment for anyone who thinks that science should deliver a description of reality that is unambiguously 'true', unquestioned, and fixed for all time.

And there you have it. Philosophers are content to contemplate and debate the nature of Metaphysical Reality, what purpose it might serve, and whether it has any real meaning or significance. They are philosophers, and do not concern themselves with the business of sailing aboard the Ship of Science.* In contrast, this is what scientists do. The ship ploughs back and forth on the Sea of Representation, refining metaphysical preconceptions here and gathering not-entirely-neutral empirical facts there, travelling in hope of a successful theory.

So let's see how that might work.

---

* Although philosophers of science will have some things to say about the design and building of the ship, how it might be captained, its instruments and maritime charts, and the nature of the journey back and forth.

3

SAILING ON THE SEA OF REPRESENTATION

How Scientific Theories Work (and Sometimes Don't)

We all tend to use the word 'theory' rather loosely. I have a theory about the reasons why British citizens voted by a narrow margin to leave the European Union in the referendum that was held in June 2016. I also have a theory about Donald Trump's outrageous success in the race to become the 45th President of the United States later that year. We can all agree that no matter how well reasoned they might be, these are 'just theories'.

But successful *scientific* theories are much more than this. They appear to tell us something deeply meaningful about how nature works. Theories such as Newton's system of mechanics, Darwin's theory of evolution, Einstein's special and general theories of relativity, and, of course, quantum mechanics are broadly accepted as contingently 'true' representations of reality and form the foundations on which we seek to understand how the Universe came to be and how we come to find ourselves here, able to theorize about it. Much of what we take for granted in our complex, Western scientific-technical culture depends on the reliable application of a number of scientific theories. We have good reasons to believe in them.

In a recent *New York Times* online article, cell biologist Kenneth R. Miller explained that a scientific theory 'doesn't mean a hunch or a guess. A theory is a system of explanations that ties together a whole bunch of facts. It not only explains those facts, but predicts what you ought to find from other observations and experiments.'[1]

This is all well and good, but *how* does that happen? Where does a scientific theory come from and how is it shaped through the confrontation of metaphysical preconceptions with the empirical facts? How does it gain acceptance and trust in the scientific community? And, most importantly for our purposes here, how should we *interpret* what the theory represents?

In other words, what exactly is involved in sailing the Ship of Science across the Sea of Representation?

Your first instinct might be to reach for a fairly conventional understanding of what's generally referred to as the 'scientific method'. You might think that scientists start out by gathering lots of empirical data, and then they look for patterns. The patterns might suggest the possibility that there is a cause-and-effect relationship or even a law of nature sitting beneath the surface which guides or governs what we observe.

Drawing on our metaphor, the scientists arm themselves with a cargo of empirical data and set sail for the shores of Metaphysical Reality. Here they collect those preconceptions about reality that are relevant to the data they want to understand, perhaps involving familiar concepts, such as space and time, and familiar (but invisible) entities such as photons or electrons, and what we already think we know about their behaviours and their properties.

If these are patterns in physical data, the scientists will typically use mathematics to assemble their preconceptions into a formal theoretical structure, involving space and time, mass and energy, and maybe further properties such as charge, spin, flavour, or colour. To be worthy of consideration, the new structure will provide the scientists with the connections they need.

The theory will say that when we do *this*, we get that pattern, and this is consistent with what is observed. The scientists then go further. Trusting the veracity of the structure, they figure out that when instead we choose to do *that*, then something we've never before observed (or thought to look for) should happen. They sail back across the Sea to the shores of Empirical Reality to make more observations or do some more experiments. When this something is indeed observed to happen, the theory gains greater credibility and acceptance.

Credit for this version of the scientific method, based on the *principle of induction*, is usually given to Francis Bacon who, after an illustrious career (and public disgrace), died of pneumonia caught whilst trying to preserve a chicken by stuffing it with snow. Now, we might feel pretty comfortable with this version of the scientific method, which remained unquestioned for several hundred years, but we should probably acknowledge that science doesn't actually work this way.

If the eternal, immutable laws of nature are to be built through inductive inference substantiated or suitably modified as a result of experiment or observation, then, philosophers of the early twentieth century argued, we must accept that these laws *can never be certain.*

Suppose we make a long series of observations on ravens all around the world. We observe that all ravens are black. We make use of induction to generalize this pattern into a 'law of black ravens'.* This would lead us to predict that the next raven

---

* Readers should note that this isn't the famous 'Raven paradox', devised by the positivist philosopher Carl Hempel in the 1940s. If all ravens are black, this logically implies that any (and every) non-black object is not a raven. But, whilst we wouldn't hesitate to accept the observation of another black raven as evidence in support of the law, we'd surely struggle to accept as supporting evidence the observation of a green apple.

observed (and any number of ravens that might be observed in the future) should also be black.

But then we have to admit that no matter how many observations we make, there would always remain the *possibility* of observing an exception, a raven of different colour, contradicting the law and necessitating its abandonment or revision. The probability of finding a non-black raven might be considered vanishingly small, but it could never be regarded to be zero and so we could never be certain that the law of black ravens would hold for all future observations. This is a conclusion reinforced by the experiences of European explorers—who might have formulated a similar 'law of white swans'—until they observed *black* swans (*Cygnus atratus*) on the lakes and rivers of Australia and New Zealand.

The philosopher Bertrand Russell put it this way: 'The man who has fed the chicken every day throughout its life at last wrings its neck instead, showing that more refined views as to the uniformity of nature would have been useful to the chicken.'[2]

Karl Popper argued that these problems are insurmountable, and so rejected induction altogether as a basis on which to build a scientific method. In two key works, *The Logic of Scientific Discovery* (first published in German in 1934) and *Conjectures and Refutations* (published in 1963), he argued that science instead proceeds through the invention of creative hypotheses, from which empirical consequences are then *deduced*. Conscious of some outstanding problems or gaps in our understanding of nature, scientists start out not with lots of empirical data, but by paying a visit to the shores of Metaphysical Reality. They conjure some bold (and, sometimes, outrageous) ideas for how nature *might* be, and then deduce the consequences on the voyage back across the Sea of Representation. The result is a formal theory.

What then happens when we reach the shores of Empirical Reality? This is easy: the theory is *tested*. It is exposed to the hard, brutal, and unforgiving facts. If the theory is not *falsified* by the data, by even just a single instance (observation of a single black swan will falsify the 'law of white swans'), then Popper argued that it remains a relevant and useful scientific theory.

We can go a bit further than Popper and suggest that if the test is based on existing data, on empirical facts we already know, then the theory will be tentatively accepted if it provides *a better explanation* of these facts than any available alternative. Better still, if the theory makes *predictions* that can be tested, and which are then upheld by data from new observations or experiments, then the theory is likely to be more widely embraced.*

I honestly doubt that many practising scientists would want to take issue with any of this. I could provide you with many, many examples from the history of science which show that this is, more or less, how it has worked out. This isn't a book about history, however, so I'll restrict myself to just one very pertinent example.

We saw in Chapter 1 that the real quantum revolutionary was Einstein, who published his light-quantum hypothesis in his 'miracle year' of 1905. It's important to note that Einstein didn't induct this hypothesis from any available data. He simply perceived a problem with the way that science was coming to understand matter—in terms of discrete atoms and molecules—and electromagnetic radiation, which was understood exclusively in terms of continuous waves. The prevailing scientific description of matter and light didn't fit with Einstein's metaphysical preconceptions about how nature ought to be. Whilst on his visit to the shores of Metaphysical Reality, he conjectured that Planck's

---

* Not surprisingly, the more counterintuitive the prediction, the more scientists are likely to look twice at where the prediction comes from. It goes something like this: That's ridiculous—how could that possibly be true? What? It *is* true? OMG! What's the theory again?

conclusions should be interpreted as though light itself is composed of discrete quanta.

Einstein then sailed across the sea, representing his light-quanta in a theory that predicted some consequences for the photoelectric effect. The rest, as they say, is history.

The light-quantum hypothesis passed the test, but as we've seen, it still remained controversial (scientists can be very stubborn). Just a few years after the experiments on the photoelectric effect, Arthur Compton and Pieter Debye showed that light could be 'bounced' off electrons, with a predictable change in the frequency (and hence the energy) of the light. These experiments demonstrated that light does indeed consist of particles moving like small projectiles. Gradually, light-quanta became less controversial and more acceptable.

Popper's take on science is known as the 'hypothetico-deductive' method. This is a bit of a clumsy term, but essentially it means that scientists draw on all their metaphysical preconceptions to hypothesize (or 'make a guess') about how nature works, and then deduce a formal theory which tells what we might expect to find in empirical observations or experiments. Science then proceeds through the confrontation between theory and the facts, as the ship sails between the shores. This is not a one-way journey—the ship makes many journeys back and forth, and those relevant metaphysical preconceptions that survive become tightly and inextricably bound into the theory (and, as we've seen, into the empirical observations, too). In this way the relevant metaphysics becomes 'naturalized' or 'habitual', justified through the success of the resulting theory.[3]

It's worth mentioning in passing that the preconceptions, data, and indeed the ship itself is conditioned by their historical and cultural contexts, or perspectives. There are passages in Newton's *Principles of Natural Philosophy*, published in 1687, that refer to God's role in keeping the stars apart, a metaphysical pre-conception that would be unusual in today's science. 'Journeys are always perspectival,' contemporary philosopher Michela Massimi told me in a discussion based on my metaphor, 'we sail our ship using the only instruments (compass and whatever else) that our current technology, theories, and experimental resources afford. So any back and forth between the shores of Empirical Reality and metaphysical posits is guided and channelled by who we are, and most importantly, by our scientific history.'[4] Compare a seventeenth-century tall ship in full sail with a modern ocean liner.

There is in principle no constraint on the nature of the hypotheses that scientists might come up with during their frequent visits to the shores of Metaphysical Reality. How, you might then ask, is science any different from any other kind of wild speculation? If I propose that some mysterious force of nature governs our daily lives depending on our birth signs, surely we would all agree that this is not science? What if I propose that *similia similibus curentur*—like cures like—and that when diluted by a factor of 10^{12} (or 10^{60}), the substances that cause human diseases provide effective cures for them? Is this science? Or is it snake oil? What if I reach for the ultimate metaphysical preconception and theorize that God is the intelligent cause of all life on planet Earth, and all that we erroneously claim to be the result of evolution by natural selection is actually God's grand design?

Your instinct might be to dismiss astrology, homeopathy, and intelligent design as pseudoscience, at best. But why? After all, they involve hypotheses based on metaphysics, from which certain theoretical principles are deduced, and they arguably make predictions which can be subjected to empirical test. We can see immediately that, given the fundamental role of metaphysics in scientific theorizing, if we are to draw a line between theories that we regard to be scientific and pseudoscience or pure meta-physics, then we need something more. We need a *demarcation criterion*.

The logical positivists proposed to use 'verification' to serve this purpose. If a theory is capable in principle of being verified through observational or experimental tests, then it can be considered to be scientific. But the principle of induction was also central to the positivists' programme and, in rejecting induction, Popper had no alternative but to reject verification as well. Logically, if induction gives no guarantees about the uniformity of nature (as Russell's chicken can attest), then the continued verification of theories gives none either. Theories tend to be verified until, one day, they're not.

As we saw earlier, Popper argued that what distinguishes a scientific theory from pseudoscience and pure metaphysics is the potential for it to be *falsified* on exposure to the empirical data. In other words, a theory is scientific if it has the *potential to be proved wrong*.

This shift is rather subtle, but it is very effective. Astrology makes predictions, but these are intentionally general, and wide open to interpretation. Popper wrote: 'It is a typical soothsayers' trick to predict things so vaguely that the predictions can hardly fail: that they become irrefutable.'[5] If, when confronted with contrary and potentially falsifying evidence, the astrologer can simply reinterpret the prediction, then this is not scientific. We can find many ways to criticize the premises of homeopathy and dismiss it as pseudoscience, as it has little or no foundation in our current understanding of Western, evidence-based medi-cine—as a theory it doesn't stand up to scrutiny. But even if we take it at face value we should admit that it fails all the tests—there is no evidence from clinical trials for the effectiveness of homeopathic remedies beyond a placebo effect. Those who stub-bornly argue for its efficacy are not doing science.

And, no matter how much we might want to believe that God designed all life on Earth, we must accept that intelligent design makes no testable predictions of its own. It is simply a conceptual alternative to evolution as the cause of life's incredible complexity. Intelligent design cannot be falsified, just as nobody can prove the existence or non-existence of God. Intelligent design is not a scientific theory: it is simply overwhelmed by its metaphysical content.

Alas, this is still not the whole story. This was perhaps always going to be a little too good to be true. The lessons from history teach us that science is profoundly messier than a simple demarcation criterion can admit. Science is, after all, a fundamentally human endeavour, and humans can be rather unpredictable things. Although there are many examples of falsified or failed scientific theories through history, science doesn't progress through an endless process of falsification. To take one example: when Newton's classical mechanics and theory of universal gravitation were used to predict the orbit of a newly discovered planet called Uranus in 1781, the prediction was found to be wrong. But this was not taken as a sign that the structures of classical mechanics and gravitation had failed.

Remember that it's actually impossible to do science without metaphysics, without some things we're obliged to accept at face value without proof. Scientific theories are constructed from abstract mathematical concepts, such as point-particles or gravitating bodies treated as though all their mass is concentrated at their centres. If we think about how Newton's laws are actually applied to practical situations, such as the calculation of planetary orbits, then we are forced to admit that no application is possible without a whole series of so-called *auxiliary* assumptions or hypotheses.

Some of these assumptions are stated, but most are implied. Obviously, if we apply Newton's mechanics to planets in the

Solar System then, among other things, we assume our knowledge of the Solar System is complete and there is no interference from the rest of the Universe. In his recent book *Time Reborn*, contemporary theorist Lee Smolin wrote: 'The method of restricting attention to a small part of the universe has enabled the success of physics from the time of Galileo. I call it *doing physics in a box*.'[6]

One of the consequences of doing physics in a box is that when predictions are falsified by the empirical evidence, it's never clear why. It might be that the theory is false, but it could simply be that one or more of the auxiliary assumptions is invalid. The evidence doesn't tell us which. This is the *Duhem–Quine thesis*, named for physicist and philosopher Pierre Duhem and philosopher Willard Van Orman Quine.

And, indeed, the problem with the orbit of Uranus was traced to one of the auxiliary assumptions. It was solved simply by making the box a little bigger. John Adams and Urbain Le Verrier independently proposed that there was an as-yet unobserved eighth planet in the Solar System that was perturbing the orbit of Uranus. In 1846 Johann Galle discovered the new planet, subsequently called Neptune, less than one degree from its predicted position.

Emboldened by his success, in 1859 Le Verrier attempted to use the same logic to solve another astronomical problem. The planetary orbits are not exact ellipses. If they were, each planet's point of closest approach to the Sun (called the perihelion) would be fixed, the planet always passing through the same point in each and every orbit. But astronomical observations had shown that with each orbit the perihelion shifts slightly, or precesses. It was understood that much of the observed precession is caused by the cumulative gravitational pull of all the other planets in the Solar System, effects which can be predicted using Newton's gravitation.

But, for the planet Mercury, lying closest to the Sun, this 'Newtonian precession' is predicted to be 532 arc-seconds per century.* The observed precession is rather more, about 575 arc-seconds per century, a difference of 43 arc-seconds. Though small, this difference accumulates and is equivalent to one 'extra' orbit every three million years or so.

Le Verrier proposed the existence of another planet, closer to the Sun than Mercury, which became known as Vulcan. Astronomers searched for it in vain. Einstein was delighted to discover that his general theory of relativity predicts a further 'relativistic' contribution of 43 arc-seconds per century, due to the curvature of spacetime around the Sun in the vicinity of Mercury.† This discovery gave Einstein the strongest emotional experience of his life in science: 'I was beside myself with joy and excitement for days.'[7]

It seems from this story that a theory is only going to be abandoned when a demonstrably *better theory* is available to replace it. We could conclude from this that scientific theories are *never falsified*, as such, they are just eventually shown to be inferior when compared with competing alternatives. Even then, demonstrably falsified theories can live on. We know that Newton's laws of motion are inferior to quantum mechanics in the microscopic realm of molecules, atoms, and subatomic particles, and they break down when stuff of any size moves at or close to the speed of light. We know that Newton's law of universal gravitation is inferior to Einstein's general theory of relativity. And yet Newton's laws remain perfectly satisfactory when applied to 'everyday'

---

* A full circle is 360°, and an arc-minute is one-sixtieth of one degree. An arc-second is then one-sixtieth of an arc-minute. So, 532 arc-seconds represents about 0.15 of a degree.

† The perihelia of other planets are also susceptible to precession caused by the curvature of spacetime, but as these planets are further away from the Sun the contributions are much less pronounced.

objects and situations and physicists and engineers will happily make use of them, even though we know they're 'not true'.

Problems like these were judged by philosophers of science to be insurmountable, and consequently Popper's falsifiability criterion was abandoned (though, curiously, it still lives on in the minds of many practising scientists). Its demise led Paul Feyerabend—something of a Loki among philosophers of science—to reject the notion of the scientific method altogether and promote an anarchistic interpretation of scientific progress. In science, he argued, *anything goes*. He encouraged scientists[8]

> to step outside the circle and either to invent a new conceptual system, for example a new theory, that clashes with the most carefully established observational results and confounds the most plausible theoretical principles, or to import such a system from outside science, from religion, from mythology, from the ideas of incompetents, or the ramblings of madmen.

According to Feyerabend, science progresses in an entirely subjective manner, and scientists should be afforded no special authority: in terms of its application of logic and reasoning, science is no different from any other form of rational inquiry. He argued that a demarcation criterion is all about putting science on a pedestal, and ultimately stifles progress as science becomes more ideological and dogmatic.

In 1983, philosopher Larry Laudan declared that the demarcation problem is intractable, and therefore a pseudo-problem.* He argued that the real distinction is between knowledge that is reliable and that which is unreliable, irrespective of its provenance. Terms like 'pseudoscience' and 'unscientific' are 'just hollow phrases which only do emotive work for us'.[9]

---

* This is something of a standard philosophical ploy.

Okay—time for some confessions. I don't buy the idea that science is fundamentally anarchic, and that it has no rules. I can accept that there are no rules associated with the creative processes that take place along the shores of Metaphysical Reality. You prefer induction? Go for it. You think it's better to deduce hypotheses and then test them? Great. Although I would personally draw the line at seeking new ideas from religion, mythology, incompetents, and madmen,* at the end of the day nobody cares overmuch how new theoretical concepts or structures are arrived at if they result in a theory that *works*.

But, for me at least, there *has* to be a difference between science and pseudoscience, and between science and pure metaphysics. As evolutionary-biologist-turned-philosopher Massimo Pigliucci has argued, 'it is high time that philosophers get their hands dirty and join the fray to make their own distinctive contributions to the all-important—and sometimes vital—distinction between sense and nonsense.'[10]

So, if we can't make use of falsifiability as a demarcation criterion, what do we use instead? I don't think we have any real alternative but to adopt what I might call the *empirical criterion*. Demarcation is not some kind of binary yes-or-no, right-or-wrong, black-or-white judgement. We have to admit shades of grey. Popper himself (who was no slouch, by the way) was more than happy to accept this:[11]

the criterion of demarcation cannot be an absolutely sharp one but will itself have degrees. There will be well-testable theories, hardly testable theories, and non-testable theories. Those which are non-testable are of no interest to empirical scientists. *They may be described as metaphysical.*

---

* Not because I necessarily have anything against religion, mythology, incompetents, and madmen as sources of potential scientific hypotheses, but because I seriously doubt the efficacy of such an approach.

Some scientists and philosophers have argued that 'testability' is to all intents and purposes equivalent to falsifiability, but I disagree. Testability implies only that the theory either make contact or, at the very least, hold some promise of making contact, with empirical evidence. It makes absolutely no presumptions about what we might actually *do* in light of the evidence. If the evidence verifies the theory, that's great—we celebrate and then start looking for another test. If the evidence fails to support the theory, then we might ponder for a while or tinker with the auxiliary assumptions. Either way, we have something to work with. *This is science.*

Returning to my grand metaphor, a well-testable theory is one for which the passage back across the sea to Empirical Reality is relatively straightforward. A hardly testable theory is one for which the passage is for whatever reason more fraught. Some theories take time to develop properly, and may even be perceived to fail if subjected to tests before their concepts and limits of applicability are fully understood. Sometimes a theory will require an all-important piece of evidence which may take time to uncover. Peter Higgs proposed the mechanism that would be named for him in a paper published in 1964, and the Higgs mechanism went on to become an essential ingredient in the standard model of particle physics, the currently accepted quantum description of all known elementary particles. But the mechanism wasn't accepted as 'true' until the tell-tale Higgs boson was discovered at the Large Hadron Collider, nearly fifty years later.

Make no mistake, if the theory fails to provide even the promise of passage across the sea—if it is trapped in the tidal forces of the whirlpool of Charybdis—then this is a non-testable theory. No matter how hard we try, we simply can't bring it back to Empirical Reality. This implies that the theory makes no predictions, or makes predictions that are vague and endlessly adjustable, more typical of the soothsayer or the snake oil salesman. This is pure metaphysics, not science, and brings me to my second-favourite Einstein quote: 'Time and again the passion for understanding has led to the illusion that man is able to comprehend the objective world rationally by pure thought without any empirical foundations—in short, by metaphysics.'[12]

I want to be absolutely clear. I've argued that it is impossible to do science of any kind without involving metaphysics in some form. The scientists' metaphysical preconceptions are essential, undeniable components in the construction of any scientific theory. But there *must* be some kind of connection with empirical evidence. There must be a tension between the ideas and the facts. The problem is not metaphysics per se but rather the nature and extent of the metaphysical content of a theory. Problems arise when *the metaphysics is all there is.*

Of course, this is just my opinion. If we accept the need for a demarcation criterion, then we should probably ask who should be responsible for using it in making judgements. Philosophers Don Ross, James Ladyman, and David Spurrett argue that individuals (like me, or them) are not best placed to make such judgements, and we should instead rely on the *institutions* of modern science.[13] These institutions impose norms and standards and provide sense-checks and error filters that should, in principle, exclude claims to objective knowledge derived from pure metaphysics. They do this simply by not funding research proposals that don't meet the criteria, or by not publishing papers in recognized scientific journals.

But, I would argue, even institutions are fallible and, like all communities, the scientific community can fall prey to groupthink.[14] We happen to be living in a time characterized by a veritable cornucopia of metaphysical preconceptions coupled with a dearth of empirical facts. We are ideas-rich, but data-poor. As we will see in Chapter 10, I personally believe the demarcation line has been crossed by a few theorists, some with strong public profiles, and I'm not entirely alone in this belief. But, at least for now, the institutions of science appear to be paying no attention.

So I encourage you to form your own opinions.

An accepted scientific theory serves at least two purposes. If it is a theory expressed in the form of one or more mathematical equations, then these equations allow us to calculate what will happen given a specific set of circumstances or inputs. We plug some numbers in, crank the handle, and we get more numbers out. The outputs might represent predictions for observations or experiments that we can then design and carry out. Or they might be useful in making a forecast, or designing a new electronic device, building a skyscraper, or planning a town's electricity network. Used in this way, our principal concerns rest with the inputs and the outputs, and we might not need to think too much about what the theory actually says. Provided we can trust its accuracy and precision, we can quite happily use the theory as a 'black box', as an instrument.

The second purpose is concerned with how the theory should be *interpreted*. The equations are expressed using a collection of concepts represented by sometimes rather abstract symbols. These concepts and symbols may represent the properties and behaviours of invisible entities such as electrons, and the strengths of forces that act on them or that they produce or carry. Most likely, the equations are structured such that everything we're interested in takes place within a three-dimensional space and in a time interval stretching from then until now, or from now until sometime in the future. The interpretation of these symbols then tells us something meaningful about the things we find in nature. This is no longer about our ability to *use* the theory; it is about how the theory informs our *understanding* of the world.

You might think I'm labouring this point. After all, isn't it rather obvious how a physical theory should be interpreted? What's the big deal? If we're prepared to accept the existence of objective reality (Realist Proposition #1), and the reality of invisible entities such as photons and electrons (Realist Proposition #2), it surely doesn't require a great leap of imagination to accept:

Realist Proposition #3: *The base concepts appearing in scientific theories represent the real properties and behaviours of real physical things.*

By 'base concepts' I mean the familiar terms we use to describe the properties and behaviours of the objects of our theories. These are concepts such as mass, momentum, energy, spin, and electric charge, with events unfolding in space and time. It's important to distinguish these from other, more abstract mathematical constructions that scientists use to manipulate the base concepts in order to perform certain types of calculations. For example, in classical mechanics it is possible to represent the complex motions of a large collection of objects more conveniently as the motion of a single point in something called configuration space or phase space. Nobody is suggesting that such abstractions should be interpreted realistically, just the base concepts that underpin them. We'll soon see that arguments about the interpretation of quantum mechanics essentially hinge on the interpretation of the wavefunction. Is the wavefunction a base concept, with real properties and real behaviours? Or is it an abstract construction?

Proposition #3 appears straightforward, but then I feel obliged to point out that the conjunction of everyday experience and familiarity with classical mechanics has blinded us to the difference between the physical world and the ways in which we choose to *represent* it.

Let's use Newton's second law of motion as a case in point. We know this law today by the rather simple equation:

force   mass   acceleration

This says that accelerated motion will result if we apply a force (or an 'action' of some kind) on an object with a certain mass. Now, whilst it is certainly true to say that the notion of mechanical force still has much relevance today, as I explained in Chapter 1 the attentions of eighteenth- and nineteenth-century physicists switched from force to *energy* as the more fundamental concept. My foot connects with a stone, this action impressing a force on the stone. But a better way of thinking about this is to see the action as transferring energy to the stone.

The term 'energy' was first introduced in the early nineteenth century and it gradually became clear that this is a conserved quantity—energy can be neither created nor destroyed and is simply moved around a physical system, shifting from one object to another or converting from one form to another. It was realized that kinetic energy—the energy associated with motion—is not in itself conserved. Physicists recognized that a system might also possess *potential* energy by virtue of its physical characteristics and situation. Once this was understood it became possible to formulate a law of conservation of the *total* energy—kinetic plus potential—and this was achieved largely through the efforts of physicists concerned with the principles of thermodynamics.

Today, we replace Newton's force with the rate of change of potential energy in space. Think about it this way. Sisyphus, the king of Ephyra, is condemned for all eternity to push an enormous boulder to the top of a hill, only for it roll back to the bottom

(Figure 8). As he pushes the boulder upwards, he expends kinetic energy on the way, determined by the mass of the boulder and the speed with which he rolls or carries it. If we neglect any losses due to friction, all this kinetic energy is transferred into potential energy, held by the boulder, perched at the top. This potential energy is represented by the way the hill slopes downwards. As the boulder rolls back down the slope, the potential energy is converted back into the kinetic energy of motion. For a given mass, the steeper the slope (the greater the force), the greater the resulting acceleration.

With this in mind, why would we hesitate, even for an instant, to accept Realist Proposition #3? Sisyphus is a figure from Greek

Figure 8 The myth of Sisyphus (painting by Titian, 1549) demonstrates the relationship between kinetic and potential energy.

mythology, but there are many real hills, and many real boulders, and we don't doubt what will happen as the boulder rolls down. Acceleration is something we've experienced many thousands of times—there is no doubting its reality.

But force = mass × acceleration is an equation of a classical scientific theory, and we must remember that it is impossible to do science without metaphysics, without assuming some things for which we can't contrive any evidence. And, lest we forget, remember that to apply this equation we must also do physics in a box.

The first thing we can acknowledge is that the slope of the hill represents the rate of change of potential energy *in space*. Acceleration is the rate of change of velocity *with time*. And, for that matter, velocity itself is the rate of change of the boulder's position *in space with time*. That Newton's second law requires space and time should come as no real surprise—it's about motion, after all.

But, as I explained in Chapter 2, Newton's absolute space and time are entirely metaphysical. Despite superficial appearances, we only ever perceive objects to be moving towards or away from each other, changing their *relative* positions. This is relative motion, occurring in a space and time that are in principle defined only by the relationships between the objects themselves. Newton's arch-rival, the philosopher Gottfried Wilhelm Leibniz, argued: 'the fiction of a finite material universe, the whole of which moves about in an infinite empty space, cannot be admitted. It is altogether unreasonable and impracticable.'[15]

Newton understood very well what he was getting himself into. So why, then, did he insist on a system of absolute space and time? Because by adopting this metaphysical preconception he found that he could formulate some very highly successful laws of motion. Success breeds a certain degree of comfort, and a willingness to suspend disbelief in the grand but sometimes rather questionable foundations on which theoretical descriptions are constructed.

Then there's the question of Newton's definition of mass. Here it is: '*The quantity of matter is the measure of the same, arising from its density and bulk conjunctly*...It is this that I mean hereafter everywhere under the name body or mass.'[16] If we interpret Newton's use of the term 'bulk' to mean volume, then the mass of an object is simply its density (say in grams per cubic centimetre) multiplied by its volume (in cubic centimetres). It doesn't take long to figure out that this definition is entirely circular, as Mach pointed out many years later: 'As we can only define density as the mass of a unit of volume, the circle is manifest.'[17]

I don't want to alarm you unduly, but no matter how real the concept of mass might seem, the truth is that we've never really understood it. Einstein messed things up considerably with his famous equation $E = mc^2$, which is more deeply meaningful when written as $m = E/c^2$: 'The mass of a body is a measure of its energy content.'[18] In the standard model of particle physics, elementary particles such as electrons are assumed to 'gain mass' through their interactions with the Higgs field (this is the Higgs mechanism). The masses of protons and neutrons (and hence the masses of all the atoms in your body) are actually derived in large part from the *energy* of the colour force (carried by gluons) that binds the up and down quarks inside them.[19]

See what I mean? If an equation as simple and familiar as force = mass × acceleration is rife with conceptual problems and difficulties of interpretation, why would we assume that we can understand anything at all?

Once again we should remember that we don't actually have to understand it completely in order to use it. We know that it works (within the limits of its applicability) and we can certainly use it to calculate lots of really useful things. But the second law of motion is much more than a black box. There is a sense in which it provides genuine understanding, even though we might be unsure about the meaning of some of its principal concepts.

Of course, we know that Newton's laws have been superseded by Einstein's special and general theories of relativity, which change the way we think about space, time, and mass. But these are still classical theories (in the sense that they're not quantum theories). Take it from me that there's nothing in relativity that should shake our confidence about Realist Proposition #3. Though they're still very much on the surface, we have to accept that even the base concepts in our theories sometimes need to be understood at a somewhat deeper level. Space, time, and mass are real but our understanding of them in Newton's second law of motion is only approximate. After all, scientific theories are provisional. They're only approximately or contingently true, valid for as long as they continue to provide descriptions that are judged to be in agreement with the evidence.

Nevertheless, if I've managed to sow a few seeds of doubt then I've done my job. As we've already seen in Chapter 1, in quantum mechanics these doubts return with a startlingly cruel vengeance.

Now, I've tended to find that discussions about the interpretation of quantum mechanics can quickly get bogged down and confused on the subject of 'reality'. There's a tendency to conflate objective reality (Realist Proposition #1), the reality of 'invisible' entities like electrons (Proposition #2), and the reality of the *representation* of the properties and behaviour of these entities in scientific theories (Proposition #3). I'd be the first to admit that these propositions are not so cleanly separable, but I'd argue that there's much to be gained by considering them as such.

It would seem that Proposition #3 is contingent on the acceptance of #1 and #2. It's surely pointless to argue for a realist interpretation of a scientific representation whilst at the same time denying that stuff is real when we're not looking at it or thinking about it, and when our only evidence for it is indirect.* But, of course, acceptance of Propositions #1 and #2 *doesn't imply acceptance of #3*. We can accept #1 and #2 but still choose to reject #3. I've come to believe that the best way to appreciate the debate about the interpretation of quantum mechanics is to view this *not* as a debate about the 'nature of reality', as such, but as a debate about the realism (or otherwise) of our *representation* of reality. In essence, it's a debate about Proposition #3.

For the sake of completeness, I want to be clear that there's a little more to scientific realism than this.[20] We need to propose further that scientific theories, interpreted realistically according to #3, meet the empirical criterion: they can be tested and either confirmed as approximately or contingently true (for now) or their predictions can be shown to be false by comparison with empirical evidence. We also need to agree that when we talk about 'progress' in science, we understand that this is based on successively *more accurate representations*, the result of sailing the Ship of Science back and forth across the sea over time, refining and tightening the relationship between our metaphysical pre-conceptions and the empirical data. The philosopher Hilary Putnam wrote: 'The positive argument for realism is that it is the only philosophy that doesn't make the success of science a miracle.'[21] This is sometimes referred to as the *no miracles argument*.

This would seem to make for a relatively straightforward distinction between realism and empiricism (or anti-realism) at the level of representation, but let's not be too hasty. In seeking to

* Whether you agree with this or not, we should acknowledge that some philosophical traditions are based on the notion of rejecting (or at least remaining agnostic about) objective reality, whilst accepting that it is still possible to devise truthful representations of sensible phenomena.

resolve some of the contradictions between realism and anti-realism, the philosopher John Worrall developed a philosophy derived from a position first described by mathematician Henri Poincaré. This is called *structural realism*. According to Worrall, scientific theories tell us only about the form or *structure* of reality, but they tell us nothing about the underlying *nature* of this reality.[*]

As one theory displaces another, the mathematics might change and even the interpretation of the base concepts might change, but the structure or network of relationships between physical things is preserved. In general, a better theory will accommodate all the relationships between phenomena established through the theory it has replaced. For example, the quantum mechanical description of photons preserves all the structural relationships associated with the phenomena of diffraction and interference previously described by the wave theory of light, despite the fact that the mathematical formulations of these theories are so very different.

Structural realism comes in two flavours. There's a Kantian flavour which suggests that scientific theories are about things-as-they-appear in the form of a structure of empirical relationships. But there are, nevertheless, metaphysical things-in-themselves that are presumed to exist because, as Kant argued, there can be no appearances without anything that appears. This was largely Poincaré's position. In another, more empiricist flavour, the structural relationships are all there is and there are no things-in-themselves. This might lead us to wonder how it is possible to establish relationships if there's nothing to relate to, but perhaps this is really rather moot. Even if the things-in-themselves exist, we can still say nothing meaningful about them.

---

[*] Note that structural realism is not another distinct interpretation of quantum mechanics—the question we will be addressing in this book is whether interpretations of quantum mechanics are structurally realist.

This kind of approach makes the realist/anti-realist distinction much less straightforward. The philosopher Ian Hacking anticipated this dilemma, and reminded us that there is more to science than theoretical representation. Science has two principal aims: theory *and* experiment. Theories represent, says Hacking, and experiments intervene. We represent in order to intervene, and we intervene in the light of our representations. He wrote:[22]

> I suspect there can be no final argument for or against realism at the level of representation. When we turn from representation to intervention, to spraying niobium balls with positrons, anti-realism has less of a grip…The final arbitrator in philosophy is not how we *think* but what we *do*.

Theories come and go and, Hacking argues, intervention—experiment—is the final arbitrator on vexed questions concerning reality. As we will see in what follows, deciding whether a representation conforms to Proposition #3 can be a bit of a tricky business. In such situations, we will find it helpful to reach for a further proposition to help bring us to a conclusion. Therefore, at risk of blurring Hacking's distinction between representation and intervention, I propose to paraphrase his arguments as follows:

Realist Proposition #4: *Scientific theories provide insight and understanding, enabling us to do some things that we might otherwise not have considered or thought possible.*

I think of this as the 'active' proposition. Only by taking the representation seriously do we have a firm basis on which to act. This might take the form of a search for new empirical data, by designing, building, and performing new experiments or making new observations. This doesn't mean that it's impossible for a 'passive', anti-realist representation to engage and motivate experimentalists. But, as we will see, this happens most often because those experimentalists who care about these things tend to favour realist representations, and are generally uncomfortable with anti-realism.

I want to contrast this with the views of the anti-realist philosopher Bas van Fraassen, who in the 1970s developed a philosophy known as *constructive empiricism,* a less dogmatic descendant of logical positivism. Van Fraassen argues that scientific theories need only be 'empirically adequate'. It is sufficient that the representation accommodates all the empirical data and enables some prediction, but we should avoid getting tangled up in too much metaphysics. The representation is an instrument. It passively represents, nothing more.

This, then, is the proposition of last resort. If there are arguments both ways at the level of Proposition #3, we will seek judgement based on what the representation encourages us to *do.* If it actively represents, then we might be inclined to accept that this is a realist representation. If it passively represents, then we might consider it to be anti-realist.

Okay. That's enough of that.* Now, where were we?

---

* Readers might be disappointed that I've nowhere mentioned philosopher Thomas Kuhn's *The Structure of Scientific Revolutions,* and his notions of normal science, conducted within a *paradigm,* and paradigm-shifting extraordinary or revolutionary science. To be honest, these notions are not wholly relevant to my thesis in this book, though they are no less fascinating for that. Constraints of space preclude more than this footnote, although I'd encourage readers to consult a few of Kuhn's critics—especially Popper—in *Criticism and the Growth of Knowledge,* edited by Imre Lakatos and Alan Musgrave, published by Cambridge University Press in 1970.

4

WHEN EINSTEIN CAME DOWN TO BREAKFAST

*Because You Can't Write a Book About
Quantum Mechanics without a Chapter on the
Bohr–Einstein Debate*

A rmed with this perspective on the business of scientific theorizing, let's return to 1927.

Bohr's debates with both Schrödinger and Heisenberg prompted a period of deep introspection. As we saw in Chapter 1, Schrödinger argued for a realistic interpretation of the wavefunction, in the sense of Proposition #3. For him the wavefunction was something physically meaningful and tangible; it was something that could be easily visualized, a base concept. Heisenberg favoured a much more positivist, or anti-realist interpretation of quantum mechanics. He rejected any suggestion of some kind of underlying wave nature of matter that could be easily visualized, preferring to focus instead only on what can be *observed*, such as the lines in an atomic spectrum, and the inherent discontinuity and uncertainty that such measurements implied. Bohr hovered between these extremes, perceiving the validity of both descriptions yet puzzled by the fact that he could find no words of his own.

After some reflection, he eventually concluded that the language of classical physics, the language of waves and particles, of causality and continuity, is quite inadequate for describing quantum phenomena. And yet, as intelligent beings experiencing a classical world, this is the only language we have.

Whatever the true nature of the electron-in-itself, the behaviour it exhibits is conditioned by the kinds of experiments we choose to perform. These, by definition, are experiments requiring apparatus of classical dimensions, resulting in effects sufficiently substantial to be observed and recorded in the laboratory, perhaps in the form of tracks in a cloud chamber, or the series of spots on an exposed photographic plate which build up to an interference pattern, as we saw in Figure 4.

So, a certain kind of experiment will yield effects that we interpret, using the language of classical physics, as electron diffraction and interference. We conclude that in this experiment the electron is a wave. Another kind of experiment will yield effects which we interpret in terms of the trajectories and collisions of localized electrons. We conclude that in this experiment the electron is a particle. Bohr reasoned that these experiments are mutually exclusive. We cannot conceive an experiment to demonstrate both types of behaviour simultaneously, not because we lack the ingenuity, but because such an experiment is simply inconceivable.

What we get is a quantum world composed of shadows cast by our classical apparatus (think Plato's cave). We can see the electron's wave shadows or we can see its particle shadows. But because we are unable to construct apparatus in anything other than classical dimensions we cannot see what the electron *really is*: we can never discover anything about the electron-in-itself. What we are left to deal with is a fundamental wave–particle duality, a quantum world whose shadows are consistently different when we choose to cast them in different ways, using different classical apparatus.

Bohr sought to resolve this dilemma by declaring that these very different, mutually exclusive behaviours are not contradictory, they are instead *complementary*. They are different shadow projections of the same objectively real things-in-themselves.

So, where does this leave Bohr on Proposition #3? This is a good question. Although Bohr was infamously obscure in many of his writings on the subject, and he was much less staunchly empiricist than Heisenberg, on balance I believe it is fair to conclude that Bohr adopted a generally anti-realist interpretation of the wavefunction. Although it's a bit of a stretch to provide only one Bohr quote in support of this conclusion (especially as this is not even a direct quote), I've nevertheless always found this rather telling. He is quoted as saying:[1]

> There is no quantum world. There is only an abstract quantum physical description. It is wrong to think that the task of physics is to find out how nature is. Physics concerns what we can say about nature.

If indeed he ever said it, much has been written about Bohr's quoted use of the phrase 'There is no quantum world', as it seems to suggest that he denied the existence of an objective reality (Proposition #1). I believe this is nonsense, and entirely characteristic of a debate that oversimplifies questions concerning 'reality'. I attach much greater significance to 'Physics concerns what we can say about nature'. Bohr's quote is all about the *representation*.

This kind of anti-realism is quite subtle. What Bohr is actually saying is that we're fundamentally limited by the classical nature of our apparatus and our measurements, and the language of classical waves and particles we use to describe what we see. It's therefore pretty pointless to speculate about the reality or otherwise of elements of the 'abstract quantum physical description', including the wavefunction, as we have absolutely no way of discovering anything about them.[2]

Heisenberg was initially resistant to Bohr's notion of complementarity, as it gave equal validity to the wave description associated with his rival Schrödinger. As their debate became more bitter and personalized, Wolfgang Pauli was called to Bohr's institute in Copenhagen in early June 1927 to calm things down and broker a peace. With Pauli's help, Bohr and Heisenberg agreed to an uneasy reconciliation.

Among the key ingredients in the resulting interpretation of quantum mechanics are Bohr's notion of wave–particle complementarity, the uncertainty principle, and Born's quantum probability. It goes without saying that, as far as these physicists were concerned, by 1927 quantum mechanics was already a complete theory, and there was nothing more to be added.

What was remarkable was the zeal with which the disciples of this new quantum orthodoxy embraced and preached the new gospel. Heisenberg spoke and wrote of the 'Kopenhagener Geist der Quantentheorie'; the 'Copenhagen spirit' of quantum theory.[3] This has become known as the *Copenhagen interpretation* although, strictly speaking, there was never really a single 'interpretation' that all its advocates bought into. Like scripture, everybody had their own personal views on what it meant.

Einstein didn't like it at all.

The stage was set for a great debate about the quantum representation of reality. This commenced at the fifth Solvay congress in Brussels, part of a series of invitation-only international conferences on physics supported by the wealthy Belgian industrialist and philanthropist Ernest Solvay. This was the first time the protagonists had an opportunity to gather together, face to face. Born and Heisenberg delivered a joint lecture, declaring that quantum mechanics is a complete theory, 'whose fundamental physical and mathematical assumptions are no longer susceptible of any modification'.[4] Schrödinger then delivered a lecture on wave mechanics. And, following an interruption to allow participants the opportunity to attend a competing conference that had been organized in Paris, Bohr presented a lecture on complementarity.

Then Einstein stood to raise an objection. He was concerned by the implications of physical events which we would now interpret as the collapse of the wavefunction. Look back at Figure 5. Before measurement, the electron wavefunction is distributed across the screen, with a probability of being found in any location where the square of the wavefunction is non-zero—Figure 5a. After measurement, we learn that the electron is 'here', in a single location—Figure 5b. However, Einstein now pointed out, we also learn *simultaneously* that the electron is definitely *not* 'there', where 'there' can be any location on the screen where we might have expected to find it.

Einstein argued that this 'assumes an entirely peculiar mechanism of action at a distance, which prevents the wave continually distributed in space from producing an action in two places on the screen'.[5] This would later become widely known as 'spooky action at a distance'. The particle, which according to the wavefunction is somehow distributed over a large region of space, becomes localized instantaneously, the act of measurement appearing to change the physical state of the system far from the point where the measurement is actually recorded. Einstein felt that this kind of action at a distance violates one of the key postulates of his special theory of relativity: no physical action, or information resulting in physical action, can be communicated at a speed faster than light. Any physical process that happens instantaneously over substantial distances violates this postulate.

We should note right away that all this talk about physical action betrays the fact that these concerns are based on a *realistic* interpretation of the wavefunction, in the spirit of Proposition #3. This is not to say that Einstein wanted to ascribe reality to the wavefunction in the same way that Schrödinger did (we will see shortly that their views were quite different). But it's important to realize that, from the outset, the Bohr–Einstein debate involved a clash of *philosophical positions*. At great risk of oversimplifying, it was a confrontation between realism and anti-realism, between acceptance and rejection of Proposition #3.

What I find quite fascinating is that Einstein was attacking a position that Bohr wasn't actually defending. But by teasing out the physical *consequences* of his realist assumptions, Einstein sought to expose inconsistencies with what was fast becoming the standard or default interpretation.

As far as Bohr himself was concerned, the Copenhagen interpretation obliges us to resist the temptation to ask: *But how does nature actually do that?* Like emergency services personnel at the scene of a tragic accident, Bohr advises us to move along, as there's nothing to see here. And there lies the rub: for what is the purpose of a scientific theory if not to aid our *understanding* of the physical world? We *want* to rubberneck at reality. The only way to do this in quantum mechanics is to take the wavefunction more literally and realistically.

The discussion continued in the dining room of the Hôtel Britannique, where the conference participants were staying. Otto Stern described what happened next:[6]

Einstein came down to breakfast and expressed his misgivings about the new quantum theory, every time [he] had invented some beautiful experiment from which one saw that [the theory] did not work...Pauli and Heisenberg, who were there, did not pay much attention, 'ach was, das stimmt schon, das stimmt schon' [ah well, it will be all right, it will be all right].

Bohr, on the other hand, reflected on it with care and in the evening, at dinner, we were all together and he cleared up the matter in detail.

Einstein developed a series of hypothetical tests, or *gedankenexperiments* (thought experiments), based on the presumption of Proposition #3. These were about matters of principle; they were not meant to be taken too literally as practical experiments that could be carried out in the laboratory.

He began by attempting to show up inconsistencies in the interpretation of the uncertainty principle, but each challenge was deftly rebutted by Bohr. However, under pressure from Einstein's insistent probing, the basis of Bohr's counterarguments underwent a subtle shift. Bohr was obliged to fall back on the notion that measurements using classical apparatus are just too 'clumsy', implying limits on what can be measured, rather than limits on what we can know. This was precisely the position for which he had criticized Heisenberg earlier in the year.

In the eyes of the majority of physicists gathered in Brussels, Bohr won the day. But Einstein remained stubbornly unconvinced, and the seeds of a much more substantial challenge were sown.

'At the next meeting with Einstein at the Solvay Conference in 1930,' wrote Bohr some years later, 'our discussions took quite a dramatic turn.'[7]

Suppose, said Einstein, we build an apparatus consisting of a box which contains a clock connected to a shutter. The shutter covers a small hole in the side. We fill the box with photons and weigh it. At a predetermined and precisely known time, the clock triggers the shutter to open for a short time interval sufficient to allow a single photon to escape. The shutter closes. We reweigh the box and, from the mass difference and $E = mc^2$, we determine the precise energy of the photon that escaped. By this means, we have measured both the energy and the time interval within which a photon has been released from the box, with a precision that contradicts the energy–time uncertainty relation. This is Einstein's 'photon box' experiment.

Bohr was quite shocked, and he didn't see the solution right away. He had a sleepless night, searching for the flaw in Einstein's argument that he was convinced must exist. By breakfast the following morning he had an answer.

On the blackboard Bohr drew a rough, pseudo-realistic sketch of the apparatus that would be required to make the measurements in the way Einstein had described them (Figure 9). In this sketch the whole box is imagined to be suspended by a spring and fitted with a pointer so that its position can be read on a scale affixed to the support. A small weight is added to align the

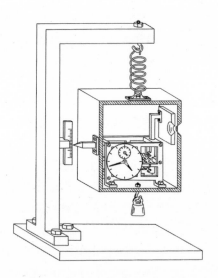

Figure 9 Einstein's 'photon box' *gedankenexperiment.*

pointer with the zero reading on the scale. The clock mechanism is shown inside the box, connected to the shutter.

After the release of one photon, the small weight is replaced by another, slightly heavier weight. This compensates for the weight lost through release of the photon so that the pointer returns to the zero of the scale. We suppose that the weight required to do this can be determined independently with unlimited precision. The difference in the two weights required to balance the box gives the mass, and hence the energy, of the photon that was released, as Einstein had argued.

So far, so good.

Bohr now drew attention to the first weighing, before the photon escapes. Obviously, the clock is set to trigger the shutter at some predetermined time and the box is sealed. We can't look at the clock because this would involve an exchange of photons—and hence energy—between the box and the outside world.

To weigh the box, a weight must be selected that sets the pointer to the zero of the scale. However, to make a precise position measurement, the pointer and scale will need to be illuminated—we need to be able to see it. But this apparatus is required to be extremely sensitive—the position of the box must change on the release of a *single* photon. So, as photons bounce off the scale, the box can be expected to jump about unpredictably. We can increase the precision of the measurement of the *average* position of the pointer by allowing ourselves a long time in which to perform the balancing procedure. This will give us the necessary precision in the weight of the box and, since we can anticipate the need for this, the clock can be set so that it opens the shutter only after this balancing procedure has been completed.

Now comes Bohr's coup de grâce.

According to Einstein's own general theory of relativity, a clock moving in a gravitational field is subject to time dilation effects. The very act of weighing a clock changes the way it keeps time. As the box bounces upwards, time slows down. As it bounces downwards, time speeds up. So, because the box is jumping about unpredictably in a gravitational field (owing to the act of balancing the weight of the box by measuring the position of the pointer), the rate of the clock is changed in a similarly unpredictable manner. This introduces an uncertainty in the exact timing of the opening of the shutter which depends on the length of time needed to complete the balancing procedure. The longer we make this procedure (the greater the ultimate precision in the measurement of the energy of the photon), the greater the uncertainty in its exact moment of release.

Bohr was able to show that the product of the uncertainties in energy and time for the photon box apparatus is entirely consistent with the uncertainty principle.

Although the photon box experiment would go on to spawn a number of research papers arguing both for and against the validity of Bohr's counterargument, Einstein conceded that Bohr's response appeared to be 'free of contradictions', but in his view it still contained 'a certain unreasonableness'.[8] At the time this was hailed as a triumph for Bohr and for the Copenhagen interpretation. Bohr had used Einstein's own general theory of relativity against him.

But note that, once again, Bohr had been obliged to defend the integrity of the uncertainty principle using arguments based on an inevitable and sizeable *disturbance* of the observed quantum system. At first sight, there seems to be no way around this. Surely, measurement of any kind will always involve interactions that are at least as big as the quantum system being measured. How can a clumsy disturbance possibly be avoided?

Einstein chose to shift the focus of his challenge. Instead of arguing that quantum mechanics—and particularly the uncertainty principle—is *inconsistent*, he now sought to derive a logical paradox arising from what he saw to be the theory's *incompleteness*. Although another five years would elapse, Bohr was quite unprepared for Einstein's next move.

Despite its seeming impossibility, Einstein needed to find a way to render Bohr's disturbance defence either irrelevant or inadmissible. This meant contriving a physical situation in which it is indeed possible in principle to acquire knowledge of the physical state of a quantum system without disturbing it in any way. Working with two young theorists, Boris Podolsky and Nathan Rosen, Einstein devised a new challenge that was extraordinarily cunning. They had found a way to do the seemingly impossible.

Imagine a situation in which two quantum particles interact and move apart. These particles may be photons, for example, emitted in rapid succession from an atom, or they could be electrons or atoms. For convenience, we'll label these particles as A and B. For our purposes, we just need to suppose that, as a result of the operation of some law of conservation, the two particles are produced in a pair of physical states that are *opposed*. It really doesn't matter what these states are, so let's just call them 'up', which we denote as ↑, and 'down', denoted ↓. So, we imagine a physical process which produces a pair of quantum particles—A and B—in ↑ and ↓ states, such that if A is ↑, B must be ↓, and if A is ↓, B must be ↑.

Here's the thing. According to quantum mechanics, the correct way to describe this kind of situation is by using a *single* wavefunction which encompasses *both* particles and *both* possible outcomes. Such a pair of particles are said to be *entangled*.

We follow the mathematical rules and write down an expression for this 'total wavefunction' which we express as a superposition of the contributions from the wavefunctions for both possible situations. In doing this we are obliged to include contributions in which A is ↑ and A is ↓, and B is ↑ and B is ↓. But our law of conservation explicitly excludes the possibility of observing pairs in which A and B are either both ↑ or both ↓. We're therefore left with something like this:

$$\text{Total} = A_\uparrow B_\downarrow + A_\downarrow B_\uparrow$$

Let's now suppose that particles A and B separate and move a large distance apart. We make a measurement on either particle to discover its state. As this is a measurement on a two-particle total wavefunction, we are obliged to represent this in terms of the expectation value of the measurement operator acting on the total wavefunction:

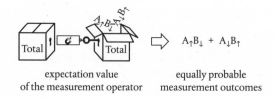

$$A_\uparrow B_\downarrow + A_\downarrow B_\uparrow$$

expectation value   equally probable
of the measurement operator   measurement outcomes

And we see that the outcomes $A_\uparrow B_\downarrow$ and $A_\downarrow B_\uparrow$ are equally probable. But, of course, for each measurement we will only ever see *one* outcome, analogous to detecting a single spot as each electron passes, one at a time, through two slits. In a realistic interpretation we must therefore presume that the total wavefunction collapses to deliver only one outcome, *either* $A_\uparrow B_\downarrow$ *or* $A_\downarrow B_\uparrow$, such that in a series of repeated measurements on identically prepared systems we will get $A_\uparrow B_\downarrow$ 50% of the time, and $A_\downarrow B_\uparrow$ 50% of the time.

Now suppose we make a measurement on A and discover that it is ↑. This *must* mean that the total wavefunction has collapsed to leave B in a ↓ state. Likewise, if we discover that A is ↓, this *must* mean that the total wavefunction has collapsed to leave B in an ↑ state. There are no other possible outcomes.

The total wavefunction relates only the probability of getting one outcome or the other, so in principle we have no way of knowing in advance whether A will be measured to be ↑ or ↓. But this really doesn't matter, for once we know the state of A, we also know the state of B with certainty, even though we may not have measured it. In other words, *we can discover the state of particle B with certainty without disturbing it in any way*. All we have to assume is that any measurement we make on particle A in no way affects or disturbs B, which could be an arbitrarily long distance away, say halfway across the Universe. We conclude that the state of particle B (and by inference, the state of particle A) *must surely have been defined all along*.

Devilish, isn't it?

In their 1935 paper, which was titled 'Can Quantum-Mechanical Description of Physical Reality be Considered Complete?', Einstein, Podolsky, and Rosen (EPR) offered a philosophically loaded definition of physical reality:[9]

> If, without in any way disturbing a system, we can predict with certainty (i.e. with a probability equal to unity) the value of a physical quantity, then there exists an element of physical reality corresponding to this physical quantity.

It's easy to see what they were trying to do. If the wavefunction is interpreted realistically, in accordance with Proposition #3, then it ought to account for the reality of the properties—the physical states of particles A and B—that it purports to describe. It clearly doesn't. There is nothing in the formulation that describes what these states are *before* we make a measurement on A, so the theory cannot be complete.

The alternative is to accept that the reality of the state of particle B is determined by the nature of a measurement we *choose* to make on a completely different particle an arbitrarily long distance away. Whatever we think might be going on, this seems to imply 'spooky action at a distance' which is at odds with the special theory of relativity. EPR argued that: 'No reasonable definition of reality could be expected to permit this.'[10]

Details of this latest challenge were reported in *The New York Times* before the EPR paper was published, in a news article titled 'Einstein Attacks Quantum Theory'. This provided a non-technical summary of the main arguments, with extensive quotations from Podolsky who, it seems, had been the principal author of the paper.

There is much in the language and nature of the arguments employed in the EPR paper that Einstein appears later to have regretted, especially the reality criterion. All the more disappointing, perhaps, as the main challenge presented by EPR does not require this (or any) criterion, though it does rest on the presumption that, whatever we make of reality, it is assumed to be *local*, meaning that as particles A and B move apart, they are assumed to exist independently of each other. Einstein deplored *The New York Times* article and the publicity surrounding it.

Nevertheless, this new challenge from Einstein sent shockwaves through the small community of quantum physicists. It hit Bohr like a 'bolt from the blue'.[11] Pauli was furious. Paul Dirac exclaimed: 'Now we have to start all over again, because Einstein proved that it does not work.'[12]

Bohr's response, when it came a short time later, inevitably targeted the reality criterion as the principal weakness. He argued that the stipulation 'without in any way disturbing a system' is essentially ambiguous, since the quantum system is influenced by the very conditions which define its future behaviour. In other words, we have to deal with elements of an *empirical reality* defined not by the quantum system in abstract, but by the quantum system in the context of the measurements we make on it and the apparatus we use. These dictate what we can expect to observe. EPR's error lies in their presumption that the wavefunction should be interpreted realistically, and the 'spooky action at a distance' is simply a consequence of this error. Just don't ask how nature actually does this, as there really is nothing to see here.

Einstein was, at least, successful in pushing Bohr to give up his clumsiness defence, and to adopt a more firmly anti-realist position. Those in the physics community who cared about these things seemed to accept that Bohr's response had put the record straight.

Schrödinger wrote to congratulate Einstein shortly after the EPR paper appeared in print. In his letter he highlighted what is, in fact, the principal challenge. When interpreted realistically, the total two-particle wavefunction is necessarily 'non-local'; it is distributed in just the same way that the electron wavefunction is distributed across the screen in the two-slit experiment. Our instinct is to imagine that, after moving a long distance apart, particles A and B are separated. They are distinct, independently existing, or 'locally real' particles. Quantum mechanics has absolutely no explanation for how we get from one situation to the other.

Einstein replied with enthusiasm, and as their correspondence continued through the summer of 1935 a further challenge to what had by now become the orthodox Copenhagen interpretation gradually emerged.

First, Einstein had to deal with Schrödinger's insistence that the wavefunction be interpreted as a description of a real 'matter wave'. Although he couldn't be clear on the details, Einstein preferred to think of the wavefunction in terms of *statistics*. We describe the properties of an atomic gas in terms of physical quantities such as temperature and pressure. But if we consider the gas as a collection of atoms, we can use the classical theories developed by Ludwig Boltzmann and James Clerk Maxwell to deduce expressions for temperature and pressure as the result of *statistical averaging* over a range of atomic motions. In this case, we deal with statistics and probabilities only because we have no way of following the motions of each individual atom in the gas. Of course, we might not be able to account for such motions except in terms of statistics, but this doesn't mean that atoms (or their motions) aren't real.

Einstein's realist interpretation of the wavefunction was very different to Schrödinger's. If quantum probability is, after all, a statistical probability born of ignorance, then there must exist a further underlying reality that we are ignorant of, just as atomic motions underlie the temperature and pressure of a gas. This was Einstein's point: as this underlying reality makes no appearance in quantum mechanics, then the theory cannot be considered to be complete. Einstein wouldn't be drawn on precisely what he thought this underlying reality might be, and the EPR paper concludes with the comment 'we left open the question of whether or not such a description exists. We believe, however, that such a theory is possible.'[13] I'll have more to say about this in a later chapter.

Schrödinger's interpretation of the wavefunction couldn't possibly be right, and Einstein sought to persuade him of this in a letter dated 8 August 1935. In this letter Einstein asked him to imagine a charge of gunpowder that, at any time over a year, may spontaneously explode. At the beginning of the year, the gunpowder is described by a wavefunction. But how should we describe the situation through the course of the year? Until we look to see what's happened, we would have to regard the wavefunction as a superposition of the wavefunctions corresponding to an explosion, and to a non-explosion. He wrote:[14]

Through no art of interpretation can this [wavefunction] be turned into an adequate description of a real state of affairs; [for] in reality there is just no intermediary between exploded and not-exploded.

Schrödinger eventually relented, and came to share Einstein's views. But the gunpowder experiment had set him thinking. As there is nothing in the mathematical formulation of quantum mechanics that accounts for the collapse of the wavefunction, then why assume this happens at the quantum level? Why not imagine that entanglement reaches all the way up the measurement chain, to the classical apparatus itself? In a reply dated 19 August he outlined another thought experiment that would become eternally enduring:[15]

Contained in a steel chamber is a Geigercounter prepared with a tiny amount of uranium, so small that in the next hour it is just as probable to expect one atomic decay as none. An amplified relay provides that the first atomic decay shatters a small bottle of prussic acid. This and—cruelly—a cat is also trapped in the steel chamber. According to the [wavefunction] for the total system, after an hour, *sit venia verbo* [pardon the phrase], the living and dead cat are smeared out in equal measure.

This is the famous paradox of Schrödinger's cat.

Einstein was in complete agreement. A total wavefunction consisting of contributions from the wavefunctions of a live and dead cat is surely a fiction. Better to try to interpret the wavefunction realistically in terms of statistics. If the experiment is duplicated, the laboratory filled with hundreds of chambers each containing a cat, then after an hour we predict that in a certain number of these (predicted as a probability derived from the wavefunction), the cat will be dead.* The Geiger counter in each

---

* And the incident will likely be followed by a visit from animal welfare authorities.

box clicks or doesn't click. If it clicks, the relay is activated, the prussic acid is released and the cat is killed. If it doesn't click, the cat survives. Nowhere in this experiment is a cat ever suspended in some kind of peculiar purgatory.

Schrödinger intended the cat paradox as a rather tongue-in-cheek dig at the apparent incompleteness of quantum mechanics, rather than a direct challenge to the Copenhagen interpretation. It does not seem to have elicited any kind of formal response from Bohr. Schrödinger wrote to Bohr on 13 October 1935 to tell him that he found his response to the challenge posed by EPR to be somewhat unsatisfactory. Surely, he argued, Bohr was over-looking the possibility that future scientific developments might undermine the assertion that the measuring apparatus must always be treated using classical physics. Bohr replied briefly that, if they were to serve as measuring instruments, then they simply could not be of quantum dimensions.

The community of physicists had in any case moved on by this time, and probably had little appetite for an endless philosophical debate that, in the view of the majority, had already been satisfactorily addressed by Bohr.

In the meantime, the 'Copenhagener Geist' had become formalized and enshrined in the very mathematical structure of quantum mechanics. The theory had emerged from a profoundly messy sequence of discoveries that had involved underlying violence to the mathematics, more than a few unjustified assumptions, and occasional conceptual leaps of faith. In its early years, the passage across the Sea of Representation had been difficult. Then there was the challenge of reconciling the two distinctly different approaches to quantum mechanics that had been developed by Heisenberg and by Schrödinger, which Schrödinger himself had demonstrated to be entirely mathematically equivalent.

In the late 1920s, Paul Dirac and John von Neumann separately sought to put the theory into some sort of order by establishing a single, formally consistent mathematical structure for quantum mechanics. Their approaches were summarized in two books: Dirac's *The Principles of Quantum Mechanics* was first published in 1930, and von Neumann's *Mathematical Foundations of Quantum Mechanics* was published in German in 1932. Their approaches were somewhat different, and von Neumann was critical of some aspects of Dirac's mathematics, but from these emerged the structure that is taught to students today.

Von Neumann had been a student of the great mathematician David Hilbert who, in a lecture delivered to the International Congress of Mathematicians in Paris in 1900, had outlined a long list of key problems that he believed would occupy the next generation of leading mathematicians. This list has become known as *Hilbert's problems*. The sixth of these concerns the mathematical treatment of physics. Hilbert argued that an important goal for future mathematicians would be to treat the physical sciences in the same manner as geometry. This means grounding physics in a set of *axioms*.

Axioms are self-evident truths that are assumed without proof, and represent the foundations of the mathematical structure that is derived from them. The proof of the axioms then lies in the consistency of the structure and the truth of the theorems that can be deduced from it. Hilbert's axiomatic method represented an almost pathological drive to eliminate any form of intuitive reasoning from the mathematics, arguing that the subject was far too important for its truths to be anything less than 'hard-wired'. Applied to physics, this demand for mathematical rigour and consistency inevitably resulted in a rather disconcerting increase in obscure symbolism and abstraction. In his review of Dirac's *Principles*, Pauli warned that Dirac's abstract formalism and focus on mathematics at the expense of physics held 'a certain danger that the theory will escape from reality'.[16] It became nearly impossible for anyone of average intelligence but without formal training in mathematics or logic fully to comprehend aspects of modern physics.

Actually, you're already quite familiar with the first few of the axioms of quantum mechanics from your reading of Chapter 1. We start with:

Axiom #1: *The state of a quantum mechanical system is completely defined by its wavefunction.*

In other words, quantum mechanics is mathematically complete, as indeed it must be if it is to serve its purpose as a foundational theory of physics. So much for Einstein. I call this the 'nothing to see here' axiom.

Axiom #2: *Observables are represented in quantum theory by a specific class of mathematical operators.*

Again, I don't propose to give you the details of what is meant here by 'specific class'. All you really need to know is that these operators are particularly suited to the task of extracting the values of observables from the wavefunction. I think of this as the 'right set of keys' axiom. To get at the observables, such as momentum and energy, we need to unlock the box represented by the wavefunction. Different observables require different keys drawn from the right set.

Axiom #3: *The average value of an observable is given by the expectation value of its corresponding operator.*

This tells us how to use the keys. I think of it as the 'open the box' axiom. It is the recipe we use to get at the observables themselves.

If quantum mechanics is to be a useful predictive theory of physics, we obviously need to know how to use it:

Axiom #4: *The probability that a measurement will yield a particular outcome is derived from the square of the corresponding wavefunction.*[*]

This is known as the 'Born rule'.[†] Or, if you prefer, you can think of this as the 'What might we get?' axiom. Note that when we apply this to a quantum superposition with two or more possible outcomes, it doesn't say what we *will* get in any individual measurement.

There is one further axiom in the main framework, related to the way we anticipate that the wavefunction will change in time:

Axiom #5: *In a closed system with no external influences, the wavefunction evolves in time according to the time-dependent Schrödinger equation.*

This means that, once established, the wavefunction evolves in a predictably deterministic and continuous manner, its properties at one moment determined entirely by its properties the

---

[*] Again, just to be clear, recall from Chapter 1 that we actually use the modulus-square of the wavefunction.

[†] Strictly speaking, the Born rule relates to the probability of finding an associated particle at a specific *position* in space. However, much the same manipulations are involved in deducing the probabilities of obtaining specific measurement outcomes from the square of the total wavefunction. What is important here is that we obtain probabilities from the squares of the wavefunctions involved, so for the sake of simplicity I will continue to call this the 'Born rule'.

moment before. Think of the electron wavefunction evolving smoothly in time as it passes through two slits, forming a 'wavefront' which alternates between high and low or zero amplitude as a result of interference, as shown in Figure 5a. This is the 'how it gets from here to there' axiom.

There is no place here for the kind of discontinuity we associate with the process of measurement. As von Neumann understood, accepting Axiom #5 forces us to adopt a further (but related) axiom in which we assume that a wavefunction representing a superposition of many measurement possibilities collapses to give a single outcome.

Of course, we never had to do anything like this in classical mechanics.

Now, Euclid's geometric axioms are concerned with the properties of straight lines, circles, and right angles and, I would contend, meet the criterion of self-evident truth. But there's nothing particularly self-evident about the axioms of quantum mechanics. I guess this is hardly surprising. The formulation of quantum mechanics is as abstract and obscure as Euclidean geometry is familiar.

These axioms leave entirely open the question of the reality or otherwise of the wavefunction—after all, this is mathematics, not philosophy. But I think it's helpful to note just how many of the Copenhagen interpretation's basic tenets became absorbed into the axiomatic structure of quantum mechanics. Just as empirical facts can never be free of some theory needed to interpret them, so theory can never be completely free of the metaphysical preconceptions that assisted at its birth. The standard mathematical formulation of quantum mechanics is not an entirely neutral witness to the debate that would follow.

This is the great game of theories. Let's now see how physicists have played this game for the past ninety years or so.

PART II
PLAYING THE GAME

5

Quantum Mechanics is Complete
SO JUST SHUT UP AND CALCULATE

The View from Scylla: The Legacy of Copenhagen,
Relational Quantum Mechanics,
and the Role of Information

The correspondence between Einstein and Schrödinger shows that, already by the summer of 1935, the Copenhagen interpretation had become the orthodoxy. It was already the default way in which physicists were meant to think about quantum mechanics. Schrödinger was pleased that EPR had 'publicly called the dogmatic quantum mechanics to account'.[1] Einstein referred to the Copenhagen interpretation as 'Talmudic'; a 'religious' philosophy that is to be interpreted only through its qualified priests, who insist on its essential truth, and who countenance no rivals.[2]

The philosopher Karl Popper called it a schism:[3]

One remarkable aspect of these discussions was the development of a split in physics. Something emerged which may be fairly described as a quantum orthodoxy: a kind of party, or school, or group, led by Niels Bohr, with the very active support of Heisenberg and Pauli; less active sympathizers were Max Born and P[ascual] Jordan and perhaps even Dirac. In other words, all the greatest names in atomic theory belonged to it, except two great men who strongly and consistently dissented: Albert Einstein and Erwin Schrödinger.

This is all really quite unfortunate. As the interpretation was forged through an uneasy alliance between Bohr and Heisenberg, brokered by Pauli, it was always going to be something of a compromise. For many physical scientists who—even today—routinely make use of quantum mechanics without worrying overmuch about what it means, the thought that there might be 'nothing to see here' is not particularly troubling. But for those (admittedly fewer) scientists who prefer to dig a little more deeply before committing to an anti-realist interpretation like this, it's fair to say that Copenhagen fails to satisfy. For every question it appears to answer, more questions go unanswered. The ambiguity and confusion it engenders, when combined with the almost pathological levels of mathematical abstraction that were first introduced into the theory by Dirac and von Neumann, render quantum mechanics virtually incomprehensible, even to many scientists.

Bohr's insistence on the principle of complementarity meant that discussions about interpretation quickly devolved into discussions about the inadequacy of the language we use to describe quantum systems and the relationship between these and the classical apparatus used to perform measurements on them. It also drew a line between the quantum and classical worlds that seems entirely arbitrary—just where is the quantum world supposed to end and the classical world begin? At the level of atoms and molecules? Or cats? The physicist John Bell called this the 'shifty split'.[4]

At the time I'm sure this must have seemed perfectly reasonable. Neither Bohr nor Heisenberg could have possibly anticipated the abilities of talented experimentalists, and the development of apparatus of extraordinary subtlety and sophistication that would be brought to bear on these questions, forty or fifty years later. As we will see in subsequent chapters, later generations of physicists would be able to make measurements on quantum systems of a kind undreamt in the philosophy of the Danish priesthood.

Complementarity and the limitations of classical apparatus— central planks in the Copenhagen interpretation—were simply not future-proof.

But these are really distractions. They are actually inessential to Bohr's core argument, which says that the quantum world is *inaccessible*. Unlike the classical world, which we had come to understand to be perfectly accessible—with its base concepts, such as momentum and energy, sitting right 'on the surface' of the equations—the quantum world lies beyond our reach. This is the principal metaphysical preconception that lurks at the heart of the Copenhagen interpretation. This preconception says that in quantum physics we have run up against a fundamental limit. We have hit the boundary that distinguishes the metaphysical things-in-themselves from the empirical things-as-they-appear.

What then happens if we forgo these additional preconceptions? What happens if we accept that our interpretation is *not* constrained by our classical language descriptions or limited by the nature of our measuring apparatus? All we need to do is separate our *experience* of quantum physics from the way we choose to *represent* it, in whatever language we deem to be appropriate. In other words, instead of being really rather ambiguous about Proposition #3, we come off the fence and reject it outright. Less Bohr, more Heisenberg.

And this is what contemporary theorist Carlo Rovelli has done.

Rovelli has spent his entire career as a theoretical physicist in pursuit of a quantum theory of gravity. In essence, this is about finding a way to bring together the two most successful foundational theories of physics—quantum mechanics and Einstein's general theory of relativity. The former describes the physics of the very small. In the form of various quantum field theories, it underpins the current standard model of particle physics. The discovery of the Higgs boson at CERN in Geneva in 2012 was only the most recent of the standard model's many triumphs. The latter is essentially a theory of space and time. It describes how mass–energy causes spacetime to curve, leading to the phenomenon we call gravity. The general theory of relativity is the basis for the current standard Big Bang model of the Universe, and the detection of gravitational waves in 2015—ripples in spacetime caused by violent events such as the merging of black holes—is only the most recent of this theory's many triumphs.

As Lee Smolin explained in his book *Three Roads to Quantum Gravity*, published in 2000, there are potentially three approaches that can be taken. You can start with quantum mechanics and seek to constrain it so that it meets the stringent requirements of general relativity. Or you can start with general relativity and find a way to 'quantize' this—yielding a theory in which space and time are themselves quantum in nature. Or you can start over, seeking a new theory which has both quantum mechanics and general relativity as limiting forms.

Rovelli and Smolin are numbered among the chief architects of a theory called *loop quantum gravity*, forged by taking the road from general relativity.[*]

---

[*] You can find out more about Rovelli, Smolin, and loop quantum gravity in Jim Baggott, *Quantum Space: Loop Quantum Gravity and the Search for the Structure of Space, Time, and the Universe*, published by Oxford University Press in 2018.

It goes without saying that if quantum mechanics is to be a foundation on which an elaborate quantum theory of gravity is to be built, a complete lack of clarity on the question of its interpretation and meaning is extremely unhelpful. Interestingly, both Rovelli and Smolin have sought clarity for themselves, but the interpretation of quantum mechanics is one of two problems in contemporary physics that they disagree on (the other is the reality of time).

In truth, whilst Rovelli has always considered quantum mechanics to be a profoundly revolutionary theory, he has never considered it to be inconsistent or incomplete in any sense. Early in their collaboration, Rovelli and Smolin were joined in deep discussion by mathematical physicist Louis Crane. They tossed ideas around about the relation between elements of reality and the observer, the philosophy of Leibniz, and quantum mechanics. 'The way I remember it,' Smolin explained, 'we each took the basic ideas and constructed theories which expressed the idea that quantum mechanics was *relational*.'[5]

In developing his theories of relativity, Einstein sought to banish from physics the entirely metaphysical concepts of absolute space and time. One consequence is that the observer is put firmly back into the picture, making measurements with rulers and clocks and performing observations *inside* the physical reality that is being examined, rather than from some unique perspective, from some kind of 'God's-eye view' of the Universe. It's worth noting that, by 'observer', we don't necessarily mean a human observer. It's enough that there be something (anything) with which a physical system can establish some kind of relationship. Instead of dealing with metaphysical things-in-themselves, we're then dealing with things-in-relation-to-other-things. If the relation is with a human observer, then we can talk about things-as-they-appear. If the relation is with some apparatus, such as a ruler or a clock, then we can talk about things-as-they-are-measured.

The observer plays a fundamentally important role in quantum mechanics, too. Just as Einstein rejected the notion that there can be absolutes in space and time, so Rovelli chose to reject the notion that a quantum system exists in an absolute, observer-independent state. In other words, in relational quantum mechanics, we can discover nothing at all about the physical quantum-states-in-themselves. He wrote: 'The thesis...is that by abandoning such a notion (in favour of the weaker notion of state—and values of physical quantities—*relative* to something), quantum mechanics makes much more sense.'[6]

In making such an assertion, Rovelli isn't rejecting the existence of an objective reality or the reality of 'invisible' entities such as electrons. He's broadly accepting of Propositions #1 and #2. There are objective, independently existing things-in-themselves—there are such things as electrons and they continue to exist when nobody's looking at them or thinking about them. But, as Kant argued, we can't discover anything about these. It only makes sense to talk about their quantum states and their properties when they establish a relationship with another system. This calls into question the viability of Proposition #3: quantum mechanics is about *relations between things,* not the real properties of real physical things independent of relation.

So, if the mathematical equations we use in quantum mechanics do not refer to the independently existing real physical states of quantum systems, to what do they then refer? Rovelli's answer is that they refer to *information* about the quantum system derived from our experience of it.

Now, Rovelli's relational interpretation does not in any way demand that we assign any special significance to the process of measurement. As far as he is concerned, measurement is just one among many different ways of establishing the relationships essential to quantum mechanics. But, of course, measurement is fundamental in that this is how we acquire knowledge about quantum systems. So let's therefore try to understand this interpretation by walking step by step through a typical measurement process.

Let's suppose, once again, that we prepare a quantum system which, based on our previous experience and understanding of the physics, can be in one of two possible states—↑ and ↓. Let's further suppose that there is only one kind of particle involved, which we label as A. As we now know, the correct way to describe this system is in terms of a total wavefunction expressed as a superposition of the contributions of the wavefunction for A in the ↑ state and the wavefunction for A in the ↓ state:

Before we go on, let's just consider where this expression has come from. We know from previous experiments that if we prepare the quantum system in just this way, our experience and understanding of the physics lead us to anticipate that it can be in one or the other state. Or, alternatively, if we establish a certain kind of *relation* between the quantum system and the device we use to prepare it, we anticipate that it can be in one or the other state. We also know that in order to make the right predictions about the future behaviour of the system, we need to represent these two states as a superposition which we call the total wavefunction. We use information from our previous experience of the physics to write the total wavefunction of the quantum system as a superposition of ↑ and ↓ states.

It's important to be clear on my use of language here. I'm using words like 'we' and 'information', which can be taken to suggest that this is once again all about human observers and information about the results of measurements which, for example, might get recorded in a laboratory notebook. And again, this is not the intended meaning. Rovelli refers to 'information' very much in a physical sense, in a form that can be manifested in inanimate objects: 'a pen on my table has information because it points in this or that direction. We do not need a human being, a cat, or a computer, to make use of this notion of information.'[7]

Now if the contributions from each of the wavefunctions in the superposition are equal, we can anticipate that in a subsequent measurement we will get *either* ↑ *or* ↓ with equal (50:50) probability. The outcomes are random: we have no way of knowing in advance which outcome we will get.

But what if, instead of measuring the ↑ or ↓ property of the system, we measure another property? Let's call this + or −. We know, again from previous experience and our understanding of the physics, that a quantum system consisting of a set of particles A prepared exclusively in the ↑ state will yield both + and − with equal probability in a subsequent measurement. Similarly, a quantum system prepared exclusively in the ↓ state will yield both + and − with equal probability.

So, what do we do now?

Remember from Chapter 1 that there is no such thing as the 'right' wavefunction. We're perfectly at liberty to choose a form for the total wavefunction that's most appropriate for the specific problem we're trying to solve. What we need is a different superposition. Instead of ↑ and ↓, we need a superposition of + and −.

And we can do this fairly easily. I won't distract you with the details. Suffice to say that we use the information from our previous experience of the behaviour of quantum systems prepared exclusively in the ↑ and ↓ states to deduce that

I'll admit this looks like we've simply substituted the wavefunctions for ↑ and ↓ with the wavefunctions for + and −. But trust me when I tell you that, no matter what it looks like, this isn't the case. There are some pretty rigorous mathematical rules that we must follow when we make this kind of change. It might help to know that the states ↑ and ↓ and + and − are often referred to as *basis states*, and what we've done therefore is *change the basis* of the representation of the total wavefunction. There is really no such thing as the 'right' or 'preferred' basis. We use information from our previous experience of the physics to change the total wavefunction to whatever basis is relevant to the problem we're looking to solve. In this case we change to a superposition of the measurement states + and −.

As before, the contributions from the wavefunctions for + and − are equal, so we can anticipate that in a subsequent measurement we will get *either* + *or* − with equal (50:50) probability. Once again, we have no way of knowing in advance which outcome we will get.

Take it from me that this is all perfectly correct. We know (again from experience) that in a series of measurements on identically prepared systems, we're likely to get a random sequence of results such as +, −, +, +, +, −, −, +,.... Although all laboratory measurements are subject to experimental errors, we also know that after making a statistically significant number of measurements, we'll find that we got + 50% of the time, and − 50% of the time.

What just happened here?

Rovelli argues that we simply use the wavefunction as a convenient way of *coding* our information about the quantum system. 'The [wavefunction] that we associate with a system...is therefore, first of all, just a coding of the outcome of these previous interactions with [the system].'[8] We do this as a way of using information derived from previous experience to make predictions for the future behaviour of the system in measurements yet to be performed. The coded information allows us to make predictions about relationships that have yet to be formed.

In other words, the wavefunction isn't real, in the sense of Proposition #3. It is not a base concept. It does not represent the real state of the quantum system. 'In [relational quantum mechanics] the quantum state is not interpreted realistically, but the position of the electron when it hits the screen is...an element of reality (although relative to the screen).'[9] The wavefunction is merely a convenient device that allows us to connect past and future.

Rejecting Proposition #3 in quantum mechanics frees us from all kinds of apparent contradictions. When we form a total wavefunction as a superposition of two possibilities, we're simply acknowledging that from previous experience we know to expect that the quantum system will produce outcomes such as ↑ or ↓, or + or −, depending on the type of measurements we're going to make. The superposition is one of information, and not real, independently existing physical states.

If the wavefunction is just coded information, then it is not required to conform to any physical laws or mechanical processes. Information isn't 'local' or 'non-local'. In itself it isn't constrained by Einstein's special theory of relativity (though any attempt to *communicate* this information will be so constrained). Information can change instantaneously. A wavefunction which consists only of information is not obliged to undergo some kind of discontinuous, physical collapse. As Rovelli explains: 'This change is unproblematic, for the same reason for which my information about China changes discontinuously any time I read an article about China in the newspaper.'[10]

This is no more mysterious than the referee tossing a coin at the beginning of a soccer or tennis match. If we felt the need, we could code the outcomes of this procedure as a superposition of 'heads' and 'tails'. The coin spins through the air and lands on the ground, and we get the result 'heads'. We believe that the two outcome possibilities persist on either side of the coin throughout, but as we're ignorant of the precise mechanics of the toss and the motion of the coin through the air we resort to probabilities. We don't tend to declare that these two possibilities 'collapse' to one outcome as the coin interacts with the ground, although we could, in principle.

In Rovelli's relational interpretation of quantum mechanics, we may understand the mechanics reasonably precisely within the limits imposed by the uncertainty principle, but we lose sight of the outcome possibilities, as we can say nothing at all about the independently existing quantum states until they have established a relation with another system. In terms of a quantum coin toss, it is as though we can anticipate the mechanical motions of the coin through the air and the number of spins it will make, but now the sides of the coin no longer exist independently except in relation to their interaction with the ground. We resort to probabilities because we're ignorant of the sides-in-themselves. We can only know the sides-on-the-ground.

Let's push this a little further. Like every material thing in the Universe, the device we use to measure the + or − state of the quantum system is also made of 'invisible' quantum objects, such as atoms consisting of quarks and electrons. Suppose we connect this device to a gauge with a readout and a pointer. If the system is measured to be in the + state, the pointer points to the left. If it is measured to be in the − state, the pointer points to the right.

What is the correct quantum-mechanical description of this situation?

Well, that depends on whether we look at the pointer. Until we take a look to see which way it's pointing, the correct representation for the total wavefunction is now something like

We've seen something like this before. The original quantum system and measuring device have become *entangled* with the gauge. Before we look to see which way the pointer went, the correct summary of the information that is available takes the form of yet another superposition.

We can go on like this forever, it seems, and this was precisely the point that Schrödinger was making with his famous cat paradox. If we rig the gauge such that pointing right kills a cat placed inside a closed box, then we've further entangled the quantum system, the original measuring device, the gauge, and the cat, giving

To discover the state of the cat we must introduce yet another device (me or you) capable of lifting the lid of the box and looking.

Our instinct is to insist that, surely, Schrödinger's cat must already be either dead or alive *before* we lift the lid. But Rovelli just shrugs his shoulders. For sure, we can *speculate* about the physical state of the cat before the 'act of measurement' but we cannot escape a simple truth: we cannot *know* the state of the cat until we establish a relationship with it, by lifting the lid, and looking.

Our mistake is to think that the superposition represents the cat's physical state—that the poor cat exists in some kind of purgatory—rather than simply representing a summary of our information about the situation. Lifting the lid doesn't collapse the wavefunction in some physical sense, dragging the cat from purgatory into a state of deadness or aliveness. There is no physical collapse. The only thing that changes when we lift the lid is the state of our knowledge of the cat and, as Rovelli says, this is unproblematic.

So let's now have a bit of fun. Alice and Bob are experimental physicists studying foundational aspects of quantum mechanics in a laboratory. Bob is running late, so Alice performs a measurement in his absence. She looks at the pointer and observes that this has moved to the left, ◨, which signals the measurement outcome A_+. She writes this down in her notebook as '+' (if she had observed the outcome ◧ she would have written '−'). We denote these results as Alice$_+$ and Alice$_-$. As far as she is concerned, the state of the quantum system is definitely +, relative to Alice. She concludes that the quantum system is now in the state given by the wavefunction

$$\left|\text{Total}\right\rangle = \left|\begin{matrix}A_+\,◨\\ \text{Alice}_+\end{matrix}\right\rangle$$

Bob is now buttonholed in the corridor by his research supervisor, who wants to know what the state of the quantum system is that Alice has just experimented on. This might seem a bit unfair, as Bob has no way of knowing, but he does know his quantum mechanics, and draws on his knowledge of the system under study (which now includes Alice) and explains that the state is given by the wavefunction

$$\left|\text{Total}\right\rangle = \left|\begin{matrix}A_+\,◨\\ \text{Alice}_+\end{matrix}\right\rangle + \left|\begin{matrix}A_-\,◧\\ \text{Alice}_-\end{matrix}\right\rangle$$

According to Bob, Alice and the result she wrote in her notebook are now entangled in the total wavefunction. He goes on to inform his supervisor that there's a 50% probability that Alice will have observed the outcome + (pointer on the left, '+' in her notebook) and a 50% probability she observed − (pointer on the right, '−' in her notebook). This might all seem perfectly reasonable—Bob can't possibly know the outcome of the measurement because he wasn't in the laboratory at the time the measurement was made. But if he now opens the door of the laboratory and asks Alice what outcome she got, then as far as Bob is concerned this constitutes a 'measurement' involving the total wavefunction in which Alice was entangled.

Before he opens the door, Alice and Bob ascribe *different states (different wavefunctions) to the quantum system*, leading Rovelli to conclude that '[i]n quantum mechanics different observers may give different accounts of the same sequence of events.'[11]

This would seem to make no sense at all if the wavefunctions are assumed to be physically real.

This logic can be extended without much difficulty to the situation envisaged by EPR, involving two entangled particles, A and B, and the quantum states ↑ and ↓. We know from Chapter 4 that the total wavefunction for such a system is given by

$$ \boxed{\Psi\ \text{Total}} = \boxed{\Psi\ A_\uparrow B_\downarrow} + \boxed{\Psi\ A_\downarrow B_\uparrow} $$

Let's presume that the two particles move apart, A moving to the left and B moving to the right. We wait until they move some long distance apart such that they are no longer in causal contact, meaning that no physical influence or information having physical consequences can pass from one to the other in the time available.* We make measurements in two separate

---

* Unless the two particles somehow communicate with each other at speeds faster than light, violating one of the fundamental postulates of Einstein's special theory of relativity (which nobody in their right mind wants to do).

laboratories.* In the laboratory over on the left, Alice observes that particle A is measured to be in an ↑ state.

Now, because she knows how the original quantum system was prepared, she can *speculate* that particle B must be in a ↓ state, but at the instant that particle A is observed she personally cannot *know* the state of B, because she hasn't established a relation with it. Likewise, in the laboratory over on the right, Bob observes that particle B is measured to be in a ↓ state, but can only speculate that particle A must therefore be ↑.

For this situation to change a *further interaction* is required which could involve Alice and Bob communicating with each other to share their results. Or perhaps they both share their results with a third observer—let's call him Charles—who concludes from this that the states of the particles are indeed correlated—A is ↑ and B is ↓. They conclude that, as a result of making a measurement on either A or B, the total wavefunction collapsed to give the outcome $A_\uparrow B_\downarrow$. They proceed to scratch their heads as they ponder on the non-locality of the total wavefunction and the spooky action at a distance implied by quantum mechanics.

But Rovelli argues that this is the wrong way to think about what's happening here. All that's really changed through this sequence is the nature of the information available to Alice, Bob, and Charles. When Alice makes her measurement, she establishes a relation with particle A. Likewise, when Bob makes his measurement, he establishes a relation with particle B which is completely independent of Alice's relation with A.

There's nothing non-local, mysterious, or spooky about any of this. But then, we might ask, how is the correlation between

---

* We'll see in Chapter 7 how real measurements in real laboratories have been performed specifically to test quantum mechanics in this way.

A and B established? That's easy. Alice and Bob used what they know about the two-particle quantum system from previous experience, and they coded this information in the total wavefunction. Remember, there's a law of conservation which means that the only possible outcomes are $A_\uparrow B_\downarrow$ and $A_\downarrow B_\uparrow$. This law specifically *excludes* the possibilities $A_\uparrow B_\uparrow$ and $A_\downarrow B_\downarrow$, which is why these were not included in the expression for the total wavefunction. In other words, information about the correlation was 'preloaded' into the total wavefunction. That the correlation is indeed observed in the two laboratories—by virtue of a further communication transmitted in a very non-spooky way at speeds no faster than light—simply reflects that *the information has been coded correctly*. All our researchers have done is take information about past events and used it to predict the outcome of a series of subsequent events.

Rovelli's interpretation requires that quantum systems enter a relationship before meaningful information about them can be gained. But there are alternatives which focus exclusively on the nature of the information associated with quantum systems. These are generally known as information-theoretic interpretations.

Drawing cues once more from Heisenberg's positivism, physicist Anton Zeilinger has suggested that quantum mechanics is essentially a theory about information, in which what we call physical properties are actually *propositions* relating to this information, derived from previous experience. Such propositions can then be determined to be true or false through future observations. 'In fact, the object therefore is a useful construct connecting observations,' Zeilinger writes.[12]

In quantum mechanics, not all propositions can be simultaneously true—'this system exhibits information characteristic of a linear particle trajectory' and 'this system exhibits information characteristic of wave interference' cannot both be true simultaneously for the same system. This means that the *amount* of information about the system is necessarily limited or constrained.

Zeilinger identifies an elementary quantum system to be one that carries information sufficient to determine the truth of just one proposition. Now quantum systems may possess a number of physical properties that are classified as polar opposites, such as positive or negative, (+ or −), up or down, (↑ or ↓), left or right. We can think of these as 'off' or 'on', as binary numbers, 0 or 1, known in computing as 'bits'. An elementary quantum system therefore carries just one bit of information. What we get now depends on what kind of experimental question we ask of the system.

Consider a quantum system prepared exclusively in an ↑ state. If we now assert the proposition 'this system exhibits information characteristic of the ↑ state', we will return the result 'true'. But what if we assert 'this system exhibits information characteristic of the + state'? As before, we're obliged to rewrite the total wavefunction of the system as a superposition of + and − states. But now the information available in the system is insufficient for a simple 'true' or 'false' pronouncement. Instead the result is completely random. In some instances it will be 'true'; in others it will be 'false', with equal (50:50) probability.

We can obviously scale up to multiple elementary systems, capable of carrying information sufficient to determine the truth of multiple propositions, involving multiple bits. An entangled state is a two-bit system involving two particles, in which the joint correlation between the particles requires more than one bit. Starting from some fairly simple information-theoretic principles, it is possible to reconstruct the entirety of quantum mechanics.

In Zeilinger's formulation, quantum information is a manifestation of the underlying quantum properties, much like temperature is a manifestation of the underlying motions of atoms and molecules. But the theorist Jeffrey Bub has argued that quantum information is a new physical 'primitive', one that cannot be reduced to physical fields or particles.[13] Bub's interpretation of information is not dependent on the existence of observers, but rather represents a fundamental element of reality itself.

As with the relational interpretation, adopting the preconception that the wavefunction represents information about a physical system and not the system itself spares us all of the uncomfortable consequences that quantum mechanics appears to imply. But at the same time it is entirely counterintuitive to separate the physics of a quantum system from our representation of it. We struggle to resist the temptation to read more into our representation than might be warranted. Rovelli sympathizes. The metaphysical preconceptions which flow from Proposition #3 are to a significant degree second nature, developed over a considerable period of acquaintance with classical mechanics. The presumption of the reality of the base concepts of our representations 'was a philosophical assumption to which science was obviously immensely indebted'.[14] But it was always an assumption.

And look at what we get if we're prepared to reject it. All our problems go away.

But the relational and information-theoretic interpretations demand a significant trade-off, and there is a heavy price to be paid. To gain these advantages we must relax our grip on reality itself, as Rovelli explains: 'the abandonment of Einstein's strict realism allows one to exempt himself from...intellectual acrobatics'.[15] We must content ourselves with what we can discover about quantum physical systems and use the mathematical formalism to interpret these in ways that allow us to predict the outcomes of future measurements. We know this works fantastically well. But don't expect this interpretation to tell us what is actually going on.

What happens *physically* to an electron on its journey from an electron gun, through a plate with two slits, to a phosphorescent screen on which it is detected as a single bright spot? What happens *physically* when a quantum system with two measurement possibilities yields just one measurement outcome? What happens *physically* to Schrödinger's cat before we lift the lid of the box, and look inside? What happens *physically* to both particles A and B when Alice detects particle A to be in an ↑ state? In this experiment, is there any kind of *physical* influence on particle B?

According to the relational and information-theoretic interpretations, there are simply no answers to these questions. This is not because we lack the wit to discover them, but because the questions themselves are meaningless. The quantum state of the electron has no significance until it establishes a relation with the phosphorescent screen, and until this happens we can *say* nothing meaningful about it. We can *say* nothing meaningful about a quantum system with two measurement possibilities until it establishes a relation with a measuring device, at which point we see one outcome. We can *say* nothing meaningful about the state of Schrödinger's cat until we lift the lid, and establish a relation with it. Irrespective of the result Alice gets for particle A, we can *say* nothing meaningful about the state of particle B until Bob establishes a relation with it. We can *say* nothing meaningful about the state of the two-particle system AB until Alice establishes a relation with A, and Bob establishes a relation with B, and they then go on to establish a relationship between themselves.

There's nothing to see here.

John Wheeler, every ready with an apt turn of phrase or a clever epithet, called it the 'great smoky dragon' (see Figure 10). We appear to have a handle on a quantum system at the start of some physical transformation—we can see the tail of the dragon—and at the finish we know the outcome—we can see the dragon's

Figure 10 Wheeler's 'great smoky dragon'.

head. But between start and finish it seems that we can say nothing meaningful about the physics. The body of the dragon is inaccessible, as though clouded in some obscure quantum fog.

Do you remember what Bohr is quoted as saying? '*It is wrong to think that the task of physics is to find out how nature is. Physics concerns what we can say about nature.*'[16]

This compares rather neatly with a quote from Alfred J. Ayer, the British spokesperson for the Vienna Circle's particular brand of positivism:[17]

> The originality of the logical positivists lay in their making the impossibility of metaphysics depend not upon the nature of what could be *known* but upon the nature of what could be *said*.

If we're also mindful of Ludwig Wittgenstein's famous caution: 'Whereof one cannot speak, thereof one must be silent',[18] we're led inexorably to a rather infamous conclusion. If we can say nothing meaningful about the physics but we have a perfectly satisfactory representation which works wonderfully well, then perhaps we should just *shut up and calculate*.

Although this last phrase is frequently attributed to Richard Feynman, it appears to have been coined by N. David Mermin. As a research student studying quantum mechanics in the 1950s, Mermin's questions about meaning and interpretation were rebuffed by his professors:[19]

> 'You'll never get a PhD if you allow yourself to be distracted by such frivolities,' they kept advising me, 'so get back to serious business and produce some results.' 'Shut up,' in other words, 'and calculate.' And so I did, and probably turned out much the better for it. At Harvard, they knew how to administer tough love in those olden days.

In my view, the relational and information-theoretic interpretations are firmly anti-realist in the sense of Proposition #3. But, mindful of Hacking's wariness of judgements based just on representation, let's ask a different question. Are the relational and information-theoretic interpretations simply passive, empirically adequate (and therefore anti-realist) interpretations, or are they something more active, in the spirit of Proposition #4?

My view is that if the wavefunction is coded information, gathered from previous experience of quantum phenomena, then these are surely rather passive representations. As we've seen, they can provide no basis for saying anything meaningful about the physics that gives rise to the information, and so they provide no deeper insight or understanding. Arguably, these approaches provide no real incentive to do anything differently, because there really is nothing to see here.

In these interpretations, we load the Ship of Science with all the empirical data we have gathered, we codify this in our passive representations, and we head straight for the rock shoal of Scylla, content with a rather empty instrumentalism.

6

Quantum Mechanics is Complete
BUT WE NEED TO REINTERPRET WHAT IT SAYS

Revisiting Quantum Probability: Reasonable
Axioms, Consistent Histories, and QBism

Where do we go from here?
Despite the implications of the Copenhagen interpretation, relational quantum mechanics, and information-theoretic
interpretations, we might still have no wish to suggest that the
quantum formalism is in any way incomplete. Is there nevertheless some solace to be gained by seeking to reinterpret what the
theory says? This offers the advantage that we avoid messing
about too much with the equations, as we know that these work
perfectly well. Instead, we look hard at what some of the symbols
in these equations might actually *mean*. This doesn't necessarily
lead us to adopt a more realist position, but it might help us to
say something more meaningful about the underlying physics
that the symbols are supposed to represent.

As a starting point, let's just acknowledge that quantum
mechanics appears to be an inherently probabilistic theory, and
that it is founded on a set of axioms. Instead of wrestling endlessly with the interpretation of the theory, is it possible to reconstruct it completely using a *different* set of axioms, in a way that
allows us to attach greater meaning to its concepts?

The theorist Lucien Hardy certainly thought so. In 2001, he posted a paper on the arXiv preprint archive in which he set out what he argued were 'five reasonable axioms' from which all of quantum mechanics can be deduced.[1] These look nothing like the axioms I presented towards the end of Chapter 4. Gone is the completeness or 'nothing to see here' axiom. Gone are the 'right set of keys' and the 'open the box' axioms. There is no assumption of the Born rule, as such.

Hardy argued that the singular feature which distinguishes quantum mechanics from any theory of physics that has gone before is indeed its probabilistic nature. So, why not forget all about wave–particle duality, wavefunctions, operators, and observables and reconstruct it as a generalized form of probability theory? In fact, the first four of Hardy's reasonable axioms serve to define the structure of *classical* probability, of the kind we would use quite happily to describe the outcomes we would anticipate from tossing a coin. It is the fifth axiom, which assumes that transformations between quantum states are continuous and reversible, which extends the foundations to include the possibility of quantum probability.* The rest of quantum mechanics then flows from these, including the Born rule.

At first sight, Hardy's fifth axiom appears rather counterintuitive. In a theory that is characterized by *discontinuities*, it seems odd to assume that transformations between quantum states happen in a smooth, continuously incremental fashion. But this is necessary to set up the kind of scenario which just can't happen in classical physics. 'Heads' can't continuously and reversibly transform into 'tails'. But the quantum states ↑ and ↓ can. Hardy's fifth axiom

---

* On further reflection, Hardy realized he could drop the demand for continuous transformation. The assumption of reversibility, when combined with his third axiom, is sufficient to necessitate continuity.

allows for the possibility of quantum superposition, entanglement, and all the fun that follows. Quantum discontinuity is then interpreted straightforwardly as the transformation of our knowledge, from two possibilities (*either* ↑ *or* ↓) to one actuality, in just the same way we see that the coin has landed with 'heads' facing up.

Hardy's paper follows something of a tradition in attempts to reconstruct quantum mechanics, and its publication sparked renewed interest in this general approach. Note, however, that any reconstruction of quantum mechanics as a general theory of probability might allow us to say some meaningful things about what goes into it and what comes out, but it tells us nothing whatsoever about what happens in between. 'What the physical system is is not specified and plays no role in the results,' explains Giulio Chiribella. Such probability theories 'are the syntax of physical theories, once we strip them of the semantics'.[2]

There's still nothing to see here.

We might be tempted to conclude that rushing to embrace an entirely probabilistic structure risks throwing the baby out with the bathwater, losing sight of whatever physics is contained within the conventional theory. Instead of discarding all the conventional axioms, is it possible just to be a little more selective?

Look back at the axioms detailed at the end of Chapter 4, and listed in the Appendix. If we're accepting of the assumption that the wavefunction provides a complete description (Axiom #1), then we need to discover how we feel about the others. There doesn't seem much to be gained by questioning the 'right set of keys', the 'open the box', or the 'how it gets from here to there' axioms, as these are most certainly necessary if we are to retain some predictability and extract the right kind of information from the wavefunction. Inevitably, our attention turns to Axiom #4, the Born rule or 'What might we get?' axiom, as this is where we sense some vulnerability.

In quantum mechanics, we tend to interpret the Born rule in terms of quantum probabilities that are established at the moment of measurement. The reason for this is quite simple and straightforward. Whether we interpret the wavefunction realistically or not, when we apply Schrödinger's wave equation we get a description of the motion that is smooth and continuous, according to Axiom #5. The form of the wavefunction at some specific time can be used to predict the form of the wavefunction at some later time. In this sense, the Schrödinger equation works in much the same way as the classical equations of motion. It is only when we introduce an interaction or a transition of some kind that changes the state of a quantum system that we're confronted with discontinuity—an electron 'jumps' to some higher-energy orbit, or the wavefunction collapses to one measurement outcome or the other, as God once more rolls the dice. This discontinuity does not—it simply cannot—appear anywhere in the Schrödinger equation.

This is the reason the Born rule is introduced as an axiom. There is nothing in the formalism itself that tells us unambiguously that this is how nature works. We apply the Born rule because this is the way we try to make sense of the inherent unpredictability of quantum physics. A quantum system has two possible measurement outcomes, but we can't predict with certainty which outcome we will get in each individual measurement. We use the Born rule in an attempt to disguise our ignorance and to pretend that we really do know what's going on. The only way we can do this is to *assume* it's true.

In conventional quantum mechanics, we assume that quantum probability arises as a direct consequence of the measurement process. But what if we don't do this? What if we reject the conventional interpretation of the Born rule, or find another, deeper, explanation for the seemingly unavoidable and inherent randomness of the quantum world?

Let's be clear. Calculating the probability of getting a particular measurement outcome from the square of the total wavefunction is deeply ingrained in the way physicists use quantum mechanics, and nobody is suggesting that this should stop. What we're suggesting instead is that the Born rule is seen not simply as a calculating device that has to be assumed because of the way quantum systems interact with our classical apparatus, but rather as an inevitable *consequence* of the underlying quantum physics, or of the way that we as human beings *perceive* this physics. Either way, this means changing the way we think about quantum probability.

We'll begin by taking a look at the first alternative.

As we've seen, the philosopher Karl Popper shared some of the realist leanings of Einstein and Schrödinger, and it is clear from his writings on quantum mechanics that he stood in direct opposition to the Copenhagen interpretation, and in particular to Heisenberg's positivism. As far as Popper was concerned, all this fuss about quantum paradoxes was the result of misconceiving the nature and role of probability.

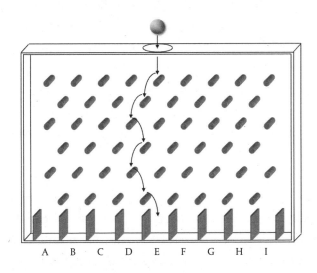

Figure 11 Popper's pin board.

To explain what he meant, Popper made extensive use of an analogy. Figure 11 shows an array of metal pins embedded in a wooden board. This is enclosed in a box with a transparent side, so that we can watch what happens when a small marble, selected so that it just fits between any two adjacent pins, is dropped into the grid from the top, as shown. On striking a pin, the marble may jump either to the left or to the right. The path followed by the marble is then determined by the sequence of random left/right jumps as it hits successive pins. We measure the position at the bottom of the grid at which the marble comes to rest.

Repeated measurements made with one marble (or with a 'beam' of identical marbles) allow us to determine the frequencies with which the individual marbles come to rest in specific channels at the bottom. As we make more and more measurements, these frequencies converge to a fixed pattern which we can interpret in terms of statistical probabilities. If successive marbles always enter the grid at precisely the same point and if the pins are identical, then we would expect a uniform distribution of probabilities, with a maximum around the centre, thinning out towards the extreme left and right. The shape of this distribution simply reflects the fact that the probability of a sequence in which there are about as many left jumps as there are right is greater than the probability of obtaining a sequence in which the marble jumps predominantly to the left or to the right.

From this, we deduce that the probability for a single marble to appear in any of the channels at the bottom (E, say) will depend on the probabilities for each left-or-right jump in the sequence. Figure 11 shows the sequence left–left–right–left–right–right, which puts the marble in the E channel. The probability for a particular measurement outcome is therefore determined by the *chain of probabilities* in each and every step in the sequence of events that gives rise to it. If we call such a sequence of events a 'history', then we note that there's more than one history in which the marble lands in the E channel. The sequences right–left–left–right–left–right and right–right–right–left–left–left will do just as well.

Popper argued that we change the *propensity* for the system to produce a particular distribution of probabilities by simply tilting the board at an angle or by removing one of the pins. He wrote:[3]

> [Removing one pin] will alter the probability for every single experiment with every single ball, *whether or not the ball actually comes near the place from which we removed the pin.*

So, here's a thought. In conventional quantum mechanics, we introduce quantum randomness at the moment of interaction, or measurement. What if, instead, the quantum world is inherently probabilistic at *all* moments? What if, just like Popper's pin board example, the probability for a specific measurement outcome reflects the chain of probabilities for events in each history that gives rise to it?

It's a little easier to think about this in the context of a more obvious quantum example. So let's return once again to our favourite quantum system (particle A), prepared in a superposition of ↑ and ↓ states. We connect our measuring device to a gauge with a pointer which moves to the left, 🔲, when A is measured to be in an ↑ state ($A_↑$), and moves to the right, 🔲, when A is measured to be in a ↓ state ($A_↓$).

In what follows, we will focus not on the wavefunction per se, but on a construction based on the wavefunction which—in terms of the pictograms we've considered so far—can be (crudely) represented like this:

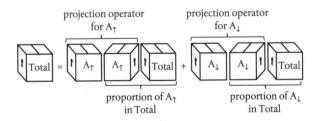

In this expression, I've applied the so-called *projection operators* derived from the wavefunctions for $A_\uparrow$ and $A_\downarrow$ onto the total wavefunction. Think of these operators as mathematical devices that allow us to 'map' the total wavefunction onto the 'space' defined by the basis functions $A_\uparrow$ and $A_\downarrow$.* If it helps, you can compare this process to that of projecting the features of the surface of the (near-spherical) Earth onto a flat, rectangular chart. The *Mercator projection* is the most familiar but this trades off the advantages of two-dimensionality for a loss of accuracy as we approach the poles, such that Greenland and Antarctica appear larger than they really are.

What happens in this quantum-mechanical projection is that the 'front end' of each projection operator combines with the total wavefunction to yield a number, which is simply related to the *proportion* of that wavefunction in the total. The 'back-end' of each projection operator is the wavefunction itself. So, what we end up with is a simple sum. The total wavefunction is given by the proportion of $A_\uparrow$ multiplied by the wavefunction for $A_\uparrow$, plus the proportion of $A_\downarrow$ multiplied by the wavefunction for $A_\downarrow$. So far in our discussions we have assumed that these proportions are equal, and we will continue to assume this.

I know that this looks like an unnecessary complication, but the projection operators take us a step closer to the actual properties ($\uparrow$ and $\downarrow$) of the quantum system, and it can be argued that these are more meaningful than the wavefunctions themselves.

How should we think about the properties of the system (represented by its projection operators) as it evolves in time through a measurement process? To simplify this some more, we'll consider just three key moments: the total (quantum plus

---

* The 'space' in question happens to be an abstract mathematical space called *Hilbert space*, named for David Hilbert.

measurement) system at some initial time shortly after preparation (we'll denote this as time t_0), the system at some later time just before the measurement takes place (t_1), and the system after measurement (t_2). What we get then is a measurement outcome that results from the sequence or 'history' of the quantum events.

Here's the interesting thing. The history we associate with conventional quantum mechanics is not the only history compatible with what we see in the laboratory, just as there are different histories that will leave the marble in the E channel of Popper's pin board.

In the *consistent histories* interpretation, first developed by physicist Robert Griffiths in 1984, these histories are organized into 'families' or what Griffiths prefers to call 'frameworks'. For the measurement process we're considering here we can devise at least three different frameworks: In Framework #1, we begin at time t_0 with an initial quantum superposition of the $A_\uparrow$ and $A_\downarrow$ states, with the measuring device (which we continue to depict as a gauge of some kind) in its 'neutral' or pre-measurement state. We suppose that as a result of some spontaneous process, by t_1 the system has evolved into either $A_\uparrow$ or $A_\downarrow$, each entangled with the gauge in its neutral state. The measurement then happens at t_2, when the gauge pointer moves either to the left or to the right, depending on which state is already present.

Time	Framework #1	Framework #2	Framework #3
t_0	$(A_\uparrow \text{ and } A_\downarrow)$⊡	$(A_\uparrow \text{ and } A_\downarrow)$⊡	$(A_\uparrow \text{ and } A_\downarrow)$⊡
t_1	$A_\uparrow$⊡ or $A_\downarrow$⊡	$A_\uparrow$⊡ and $A_\downarrow$⊡	$A_\uparrow$⊡ and $A_\downarrow$⊡
t_2	$A_\uparrow$⊡ or $A_\downarrow$⊡	$A_\uparrow$⊡ or $A_\downarrow$⊡	$A_\uparrow$⊡ and $A_\downarrow$⊡

Framework #2 is closest to how the conventional quantum formalism encourages us to think about this process. In this family of histories, the initial superposition entangles with the gauge, only separating into distinct $A_\uparrow$◨ or $A_\downarrow$◨ states at t_2, which is where we imagine or assume the 'collapse' to occur. There is no such collapse in Framework #3, in which $A_\uparrow$◨ and $A_\downarrow$◨ are entangled at t_2, producing a macroscopic quantum superposition (also known, for obvious reasons, as a Schrödinger cat state).

These different frameworks are internally consistent but mutually exclusive. We can assign probabilities to different histories *within* each framework using the Born rule, and this is what makes them consistent. But, as Griffiths explains: 'In quantum mechanics it is often the case that various incompatible frameworks exist that might be employed to discuss a particular situation, and the physicist can use any one of them, or contemplate several of them.'[4] Each provides a valid description of events, but they are distinct and they cannot be combined.

At a stroke, this interpretation renders any debate about the boundary between the quantum and classical worlds—Bell's 'shifty split'—completely irrelevant. All frameworks are equally valid, and physicists can pick and choose the framework most appropriate to the problem they're interested in. Of course, it's difficult for us to resist the temptation to ask: But what is the 'right' framework? In the consistent histories interpretation, there isn't one. Just as there is no such thing as the 'right' wavefunction, and there is no 'preferred' basis.

But doesn't the change in physical state suggested by the events happening between t_0 and t_1 in Framework #1, and between t_1 and t_2 in Framework #2, still imply a collapse of some kind? No, it doesn't:[5]

> Another way to avoid these difficulties is to think of wave function collapse not as a physical effect produced by the measuring

apparatus, but as a mathematical procedure for calculating statistical correlations…That is, 'collapse' is something which takes place in the theorist's notebook, rather than the experimentalist's laboratory.

We know by now what this implies from our earlier discussion of the relational and information-theoretic interpretations. The wavefunctions (and hence the projection operators derived from them) in the consistent histories interpretation are not real. Griffiths treats the wavefunction as a purely mathematical construct, a *pre-probability*, which enables the calculation of quantum probabilities within each framework. From this we can conclude that the consistent histories interpretation is anti-realist. It involves a rejection of Proposition #3.

The consistent histories interpretation is most powerful when we consider different kinds of questions. Think back to the two-slit interference experiment with electrons. Now suppose that we use a weak source of low-energy photons in an attempt to discover which slit each electron passes through. The photons don't throw the electron off course, but if they are scattered from one slit or the other, this signals which way the electron went. We allow the experiment to run, and as the bright spots on the phosphorescent screen accumulate, we anticipate the build-up of an interference pattern (Figure 4). In this way we reveal both particle-like, 'Which way did it go?', *and* wave-like interference behaviour at the same time.

Not so fast. In the consistent histories interpretation, it is straightforward to show that which way and interference behaviours belong to different incompatible frameworks. If we think of these alternatives as involving 'particle histories' (with 'which way' trajectories) or 'wave histories' (with interference effects), then the consistent histories interpretation is essentially a restatement of Bohr's principle of complementarity couched in the language of probability. There simply is no framework in which

both particle-like and wave-like properties can appear simultaneously. In this sense, consistent histories is not intended as an
alternative, 'but as a fully consistent and clear statement of basic
quantum mechanics, "Copenhagen done right"'.[6]

But there's a problem. If we rely on the Born rule to determine
the probabilities for different histories within each framework,
then we must acknowledge an inescapable truth of the resulting
algebra. Until it interacts with a measuring device, the square of
the total wavefunction may contain 'cross terms' or 'interference
terms':

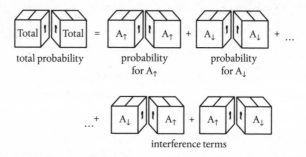

As the name implies, the interference terms are responsible
for interference, of precisely the sort that gives rise to alternating
bright and dark fringes in the two-slit experiment. In conventional quantum mechanics, the collapse of the wavefunction
implies not only a random choice between the outcomes $A_\uparrow$ and
$A_\downarrow$, but also the disappearance of the interference terms.

We can observe interference effects using light, or electrons,
or (as we'll see in Chapter 8), with large molecules or small
superconducting rings. But it goes without saying that we don't
observe interference of any kind in large, laboratory-sized
objects, such as gauge pointers or cats. So we need to find a
mechanism to account for this.

Bohr assumed the existence of a boundary between the quantum and the classical worlds, without ever being explicit about

where this might be or how it might work. But we know that any classical measuring device must be composed of quantum entities, such as atoms and molecules. We therefore expect that the first stages of an interaction between a quantum system and a classical detector are likely to be quantum in nature. We can further expect that the sheer *number* of quantum states involved quickly mushrooms as the initial interaction is amplified and converted into a signal that a human experimenter can perceive— perhaps as a bright spot on a phosphorescent screen, or the changing direction of a gauge pointer.

In the example we considered earlier, the presence in the detector of particle A in an ↑ state triggers a cascade of ever more complex interactions, with each step in the sequence governed by a probability. Although each interaction taken individually is in principle reversible, the process is quickly overwhelmed by the 'noise' and the complexity in the environment and so *appears* irreversible. Just as a smashed cocktail glass on the floor doesn't spontaneously reassemble, no matter how long we wait, though there's nothing in the classical theory of statistical mechanics that says this can't happen.

This 'washing out' of quantum interference effects as the measurement process grows in scale and complexity is called *decoherence*. In 1991, Murray Gell-Mann and James Hartle extended the consistency conditions of the consistent histories interpretation specifically to account for the suppression of interference terms through decoherence. The resulting interpretation is now more frequently referred to as *decoherent histories*.

We will meet decoherence again. But I'd like to note in passing that this is a *mechanism* for translating phenomena at the microscopic quantum scale to things we observe at our macroscopic classical scale, designed to eliminate all the strange quantum quirkiness along the way. Decoherence is deployed in a number of different interpretations, as we'll see. In this particular instance, decoherence is used as a rather general, and somewhat abstract, mathematical technique which is used to 'cleanse' the probabilities arising from the interference terms. This is entirely consistent with the view that the wavefunction is a pre-probability, and so not physically real. Other interpretations which take a more realistic view of the wavefunction make use of decoherence as a real physical process.

One last point. Decoherence rids us of the interference terms. But it does not force the choice of measurement outcome (*either* $A_\uparrow$ *or* $A_\downarrow$)—which is still left to random chance. Einstein would not have been satisfied.

Gell-Mann and Hartle were motivated in their search for an alternative to the Copenhagen interpretation as this appears to attach a special significance to the process of measurement. At the time this sat rather uncomfortably with emerging theories of quantum cosmology—in which quantum mechanics is applied to the entire Universe—because, by definition, there can in theory be nothing 'outside' the Universe to make measurements on it. The decoherent histories interpretation resolves this problem by making measurement no more significant than any other kind of quantum event.

Interest in the interpretation grew, promoted by a small but influential international group of physicists that included Griffiths, Roland Omnès, Gell-Mann, and Hartle.

But the concerns grew, too.

In 1996, theorists Fay Dowker and Adrian Kent showed that serious problems arise when the frameworks are carried through to classical scales. Whilst the history of the world with which we are familiar may indeed be a consistent history, it is not the only one admitted by the interpretation.[7] There is an infinite number of other histories, too. Because all the events within each history are probabilistic in nature, some of these histories include a familiar sequence of events but then abruptly change to an utterly unfamiliar sequence. There are histories that are classical now but which in the past were superpositions of other classical histories, suggesting that we have no basis on which to conclude that the discovery of dinosaur fossils today means that dinosaurs roamed the Earth a hundred million years ago.

Because there is no 'right' framework that emerges uniquely as a result of the exercise of some law of nature, the interpretation regards all possible frameworks to be equally valid and the choice then depends on the kinds of questions we ask. This appears to leave us with a significant context dependence, in which our ability to make sense of the physics seems to depend on our ability to ask the 'right' questions. Rather like the vast computer Deep Thought, built to answer the ultimate question of 'life, the universe and everything' in Douglas Adams's *Hitch-hiker's Guide to the Galaxy*, we are furnished with the *answer*,* but we can only hope to make sense of this if we can be more specific about the *question*.

Griffiths acknowledges that Dowker and Kent's concerns are valid, but concludes that this is the price that must be paid. The decoherent histories interpretation is[8]

> contrary to a deeply rooted faith or intuition, shared by philosophers, physicists, and the proverbial man in the street, that at any point in time there is one and only one state of the universe which is 'true', and with which every true statement about the world must be consistent. [This intuition] must be abandoned if the histories interpretation of quantum theory is on the right track.

Sometimes, in situations like this, I find it helpful to step back. Maybe I'll make another cup of tea, stop thinking about quantum mechanics for a while, and hope that my nagging headache will go away.

---

* 42.

In these moments of quiet reflection, my mind wanders (as it often does) to the state of my personal finances. Now, I regard myself as a rational person. When faced with a choice between two actions, I will tend to choose the action that maximizes some expected utility, such as my personal wealth. This seems straightforward, but the world is a complex and often unpredictable place, especially in the age of Trump and Brexit. I understood quite some time ago that buying weekly tickets for the national lottery does not make for a robust pension plan. But beyond such obvious realizations, how do any of us know which actions to take? Should I keep my money in my bank account or invest in government bonds or the stock market?

Our rationalist tendency is to put a probability on each action and choose the action with the highest probability of delivering the expected utility. We don't necessarily calculate these probabilities: we might look at bank interest rates, study the stock market, and try to form a rational, though qualitative, view. Or we might run with some largely subjective opinions about these different choices taking into account our perceptions and appetite for financial risk. Of course, we might shift the burden of responsibility for these choices to a financial adviser, if we can afford one, but we're still going to want to hear the rationale behind them before we commit.

This notion of probability as a measure of our subjective degree of belief or uncertainty is credited to the eighteenth-century statistician, philosopher, and Presbyterian minister Thomas Bayes. Bayes' approach was given its modern mathematical formulation by Pierre Simon Laplace in 1812.

Suppose you form a hypothesis that some statement might be true or valid. In Bayesian probability theory, you assign this hypothesis a probability of being valid, or as a measure of the extent of your belief in it. This is called a *prior probability*. Now look at it again in the light of some factual evidence. The probability that your hypothesis is valid in the light of the evidence is then called a *posterior probability*.

You're now faced with a simple question. Is the posterior probability larger or smaller than the prior probability? In other words, does the evidence confirm or at least support your hypothesis, does it disconfirm or serve to undermine it? Or is it neutral? Bayesian probability theory is used extensively in science, and especially in the philosophy of science as a way of thinking about how we use empirical evidence to confirm or disconfirm scientific theories.

But if we think about this for a while, we will conclude that these probabilities are really all rather *subjective*. I might come to believe one thing, but you might look at the same evidence and come to believe something completely different. Who is to say which of us is right? We are able to get away with this kind of subjectivity in daily life, but surely this has no place in theories of physics based on *objective* facts about an *objective* reality (Proposition #1).*

Perhaps this is a good point to provide a more extended version of this quote from Heisenberg:[9]

> Our actual situation in research work in atomic physics is usually this: we wish to understand a certain phenomenon, we wish to recognise how this phenomenon follows from the general laws of nature. Therefore, that part of matter or radiation which takes part in the phenomenon is the natural 'object' in the theoretical treatment and should be separated in this respect from the tools used to study the phenomenon. This again emphasises a *subjective* element in the description of atomic events, since the measuring device has been constructed by

---

* These realist propositions are handily summarized in the Appendix if you need to refer back to them.

the observer, and we have to remember that what we observe is not nature in itself but nature exposed to our method of questioning.

When scientists go about their business—observing, experimenting, theorizing, predicting, testing, and so on—they tend to do so with a certain fixed attitude or mindset. Scientists tend to assume that there is, in fact, nothing particularly special about 'us'. We are not uniquely privileged observers of the Universe we inhabit. We are not at the centre of everything. This is the 'Copernican Principle': science strives for a description in which our existence is a natural *consequence* of reality rather than the *reason* for it.

Remember that one consequence of Einstein's theories of relativity is that the observer is put back into the reality that is being observed. So, shouldn't we at least accept the need to put the experimenter back into the quantum reality that is being experimented on? We don't have to go so far as to suggest a causal connection—we don't need to reject Proposition #1 and argue that the Moon ceases to exist when nobody looks at it or thinks about it. Perhaps we just need to accept that our scientific description isn't really complete unless we place ourselves firmly in the thick of it.

In 2002, Carlton Caves, Christopher Fuchs, and Rüdiger Schack proposed to do just this. Instead of denying the subjective element in quantum mechanics they embraced it. They argued that quantum probabilities computed using the Born rule are not objective probabilities related in a mechanical way to the underlying quantum physics. They are Bayesian probabilities reflecting the personal, subjective degree of belief of the individual experimenter, related only to the experimenter's *experience* of the physics.

Your first instinct might be to reject this idea out of hand. Surely, there's a world of difference between my subjective beliefs about the stock market and the unassailably objective facts of physics? But think about it. My ability to predict movements in the stock market are limited by my lack of experience and knowledge. If I took the time to expand my experience, build my knowledge, and codify this in a couple of useful algorithms, there's a good chance I'd be able to make more realistic predictions (just ask Warren Buffet).

How is quantum physics different? For the past hundred years or so, physicists have taken the time to expand their experience, building a body of knowledge about quantum systems which is codified in the set of equations we call quantum mechanics. Why believe in the Born rule? Because this is what any rational physicist with access to the experience, knowledge, and algorithms of quantum mechanics will choose to do. 'The physical law that prescribes quantum probabilities is indeed fundamental, but the reason is that it is a fundamental rule of inference—a law of thought—for Bayesian probabilities.'[10]

This is an interpretation known as Quantum Bayesianism, abbreviated as QBism (pronounced 'cubism'). It is entirely subjective. QBists view quantum mechanics as 'an intellectual tool for helping its users interact with the world to predict, control and understand their experiences of it'.[11] Despite the exhortations of his professors, Mermin converted to QBism following six weeks in the company of Fuchs and Schack at the Stellenbosch Institute for Advanced Study in South Africa in 2012, where he 'finally began to understand what they had been trying to tell me for the past ten years'.[12]

The approach reaches beyond the Born rule to the quantum states themselves, and the wavefunctions we use to represent them. The Schrödinger equation simply constrains the way any rational physicist will choose to describe their experience, until such time as they become aware of the outcome of a measurement.

And this is unproblematic, for the same reason that Rovelli argued that his knowledge about China changes instantaneously whenever he chooses to read an article about China in the newspaper.

As the gauge pointer moves to the left, the rational Alice chooses to describe this experience of the outcome of a measurement in terms of the quantum state A_+. She expresses her degree of belief in this outcome by entering a '+' in her laboratory notebook. The rational Bob, stuck outside in the corridor with his research supervisor, chooses to describe his experiences in terms of a macroscopic quantum superposition involving the quantum system, measuring device, gauge, Alice, and her notebook. When Bob finally enters the laboratory, Alice shows him her notebook and Bob's experience and beliefs change. This is Bob's 'measurement'. It doesn't involve any quantum systems, detection devices, or gauges. Bob makes his measurement just by looking at Alice's notebook, or simply by asking her a question. Of course, Bob wasn't present when Alice made her measurement, but he trusts Alice implicitly and his degree of belief in the outcome is unshaken.

By making this all about subjective experiences, once again all the problems associated with a realist interpretation of the wavefunction evaporate. Quantum probability is a personal judgement about the physics; it says nothing about the physics itself.[13] There is no collapse of the wavefunction, for the simple reason that there are no outcomes before the act of measurement (however this is defined): experiences can't exist before they are experienced. There is no such thing as non-locality, and no spooky action at a distance: 'QBist quantum mechanics is local because its entire purpose is to enable any single agent to organize her own degrees of belief about the contents of her own personal experience. No agent can move faster than light.'[14]

This is a 'single-user' interpretation. The experiences and degrees of belief are unique to the individual—the Bayesian probabilities make no sense when applied to many individuals at once. We have to face up to the fact that the subjective nature of our individual experiences means that *we all necessarily carry different versions of reality around with us in our own minds*. If this is really the case, how is science of any kind even possible?

Calm down. The versions of reality that we all carry in our minds are still shaped by our experiences of a single, external, Empirical Reality. As a result of all our experiences, learning, and communicating with our fellow humans we develop what the philosopher John Searle calls the *background*, which I mentioned briefly in Chapter 2. This is an enormously wide and varied backdrop against which we interact with external reality. It is everything we learn from experience and come to take for granted, social and physical, as we live out our daily lives. The background is where we find all the regularities and the continuity, the expectation that the Sun will rise tomorrow, that things will be found where we left them, that cars won't turn into trees, that this $20 bill really is worth $20, and that when you turn the next page it will be covered by profoundly interesting text, and not pictures of sausages.

We each form the background by accumulating a set of mental impressions. But these have great similarity, derived from a broad set of common experiences (including experiences of quantum physics), a common body of knowledge, commonly accessible forms of communication, and human empathy. It is the close *similarity* of these individual backgrounds that makes human interaction possible.

Similar, but not the same. Within my mind is the reality with which I have learned to interact. You have no access to this reality, because you have no access to my mind. Within your mind is the reality with which you have learned to interact. I have no access to this reality, because I have no access to your mind. My reality is not your reality. But these individual realities possess many common features, such as the recognition that a \$20 bill is money, or that if I perform this experiment I'll get the result $A_\uparrow$ 50% of the time. Through the extraordinary complexity of our everyday interactions, *we perceive these separate realities as one.*

Clearly, QBism rejects Proposition #3 and in this regard is unashamedly anti-realist at the level of representation. It has nothing meaningful to say about the physics underlying the experiences. Once more, there's nothing to see here.

The Copenhagen interpretation seeks to place the blame for the inaccessibility of the quantum world on our classical language and apparatus. Rovelli's relational interpretation shifts the blame to the need to establish relationships with quantum states if they are to acquire any physical significance. Interpretations based on information do much the same. In the consistent or decoherent histories interpretation, the blame resides in the fundamentally probabilistic nature of all quantum events, and the lack of a rule to determine the 'right' framework.

In QBism, *all of physics* beyond our experience is in principle inaccessible. This kind of subjectivism applies equally well to classical mechanics, in which we codify our experience in equations that represent the behaviour of classical objects in terms of things such as mass, velocity, momentum, and acceleration.[15] Arguably, we are forced to acknowledge this subjectivism only in quantum mechanics, when we're finally confronted with the bizarre consequences of adopting a realist perspective.

But Fuchs argues that QBism is not instrumentalist. Inspired by many of John Wheeler's arguments, he prefers to think of the interpretation as involving a kind of 'participatory realism' (more on this to follow). This is participatory not in the sense of human perception and experience being necessary to conjure something from nothing and 'make it real', which would involve rejecting Propositions #1 and #2. Instead QBism simply argues that, in quantum mechanics, we can no longer ignore the fact that we are very much part of the reality we're trying so desperately hard to describe:[16]

> QBism breaks into a territory the vast majority of those declaring they have a scientific worldview would be loath to enter. And that is that the agents (observers) matter as much as electrons and atoms in the construction of the actual world—the agents using quantum theory are not incidental to it.

Hardy's axiomatic reconstruction, consistent histories, and QBism all require some substantial trade-offs. Yes, all the problems go away and we can forget about them. But we are left to contemplate a reality made of probabilities, and nothing more, or the abandonment of a single version of the historical truth, or a reality that is inherently subjective and participatory. It's clear that none of these attempts can provide us with any new insights or understanding of the underlying physics. In the context of Proposition #4 they are passive, not active, reconstructions or interpretations.

It seems that even if we're not still trapped by Scylla, mercilessly exposed to her brutally monstrous charms, then we haven't managed to sail the ship very far.

7

Quantum Mechanics is Incomplete
SO WE NEED TO ADD SOME THINGS

Statistical Interpretations Based on Local and Crypto Non-local Hidden Variables

Chapters 5 and 6 summarize interpretations of quantum mechanics that follow from the legacy of Copenhagen. They are based on a set of metaphysical preconceptions that tend to side with the anti-realism of Bohr and (especially) Heisenberg, based on the premise that quantum mechanics is complete. In the inaccessible quantum world we've finally run up against the boundary between things-in-themselves and things-as-they-appear that philosophers have been warning us about for centuries. We have to come to terms with the fact that there's nothing to see here, and we've reached the end of the road.

But this anti-realist perspective is not to everybody's taste, as theorist John Bell made all too clear in 1981:[1]

> Making a virtue of necessity, and influenced by positivistic and instrumentalist philosophies, many came to hold not only that it is difficult to find a coherent picture but that it is wrong to look for one—if not actually immoral then certainly unprofessional.

If we make the philosophical choice to side with Einstein, Schrödinger, and Popper and adopt a more realist position, this means we can't help but indulge our inner metaphysician. We can't help speculating about a reality beyond the empirical data, a reality lying beneath the things-as-they-appear. We must admit that quantum mechanics is incomplete, and be ready to make frequent visits to the shores of Metaphysical Reality in the hope of finding something—anything—that might help us to complete it.

In doing this we might be led astray, but I honestly think that it goes against the grain of human nature not to *try*.

Opening the door to realism immediately puts us right back in a fine mess. Any realist interpretation, reformulation, or extension of quantum mechanics necessarily drags along with it all the associated metaphysical preconceptions about how reality *ought* to be. It must address all the bizarre things that quantum physics appears to allow, such as superpositions now involving real physical states (rather than coded information), the instantaneous collapse of the wavefunction, and the spooky action at a distance this would seem to imply. It has to explain away or eliminate the randomness and discontinuity inherent in quantum mechanics and restore some sense of continuity and cause-and-effect, presided over by a God free of a gambling addiction. It has to find a way to make the quantum world compatible with the classical world, explaining or avoiding an arbitrary 'shifty split' between the two.

Where do we start?

In his debate with Bohr and his correspondence with Schrödinger, Einstein had hinted at a *statistical* interpretation. In his opinion, quantum probabilities, derived as the squares of the wavefunctions,* actually represent statistical probabilities, averaged over large numbers of physically real particles. We resort to probabilities because we're ignorant of the properties

---

* Actually, modulus-…okay—you've got it now, so I'll stop.

and behaviours of the physically real quantum things. This is very different from anti-realist interpretations which resort to probabilities based on previous experience because we can say nothing meaningful about any of the underlying physics.

Einstein toyed with just such an approach in May 1927. This was a modification of quantum mechanics that combined classical wave and particle descriptions, with the wavefunction taking the role of a 'guiding field' (in German, a *Führungsfeld*), guiding or 'piloting' the physically real particles. In this kind of scheme, the wavefunction is responsible for all the wave-like effects, such as diffraction and interference, but the particles maintain their integrity as localized, physically real entities. Instead of waves *or* particles, as complementarity and the Copenhagen interpretation demands, Einstein's adaptation of quantum mechanics was constructed from waves *and* particles.

But Einstein lost his enthusiasm for this approach within a matter of weeks of formulating it. It hadn't come out as he'd hoped. The wavefunction had taken on a significance much greater than merely statistical. It was almost sinister. Einstein thought the problem was that distant particles were exerting some kind of strange force on one another, which he really didn't like. But the real problem was that the guiding field is capable of exerting spooky non-local influences—changing something here instantaneously changes some other thing, a long way over there. He withdrew a paper he had written on the approach before it could be published. It survives in the Einstein Archives as a handwritten manuscript.[2]

We'll be returning to this kind of 'pilot-wave' description in Chapter 8. This experience probably led Einstein to conclude that his initial belief—that quantum mechanics could be completed through a more direct fusion of classical wave and particle concepts—was misguided. He subsequently expressed the opinion that a complete theory could only emerge from a much more radical revision of the entire theoretical structure. Quantum mechanics would eventually be replaced by an elusive grand unified field theory, the search for which took up most of Einstein's intellectual energy in the last decades of his life.

This early attempt by Einstein at completing quantum mechanics is known generally as a *hidden variables formulation*, or just a 'hidden variables theory'. It is based on the idea that there is some aspect of the physics that governs what we see in an experiment, but which makes no appearance in the representation. There are, of course, many precedents for this kind of approach in the history of science. As I've already explained, Boltzmann formulated a statistical theory of thermodynamics based on the 'hidden' motions of real atoms and molecules. Likewise, in Einstein's abortive attempt to rethink quantum mechanics, it is the positions and motions of real particles, guided by the wavefunction, that are hidden.

However, in his 1932 book *The Mathematical Foundations of Quantum Mechanics*, von Neumann presented a proof which appeared to demonstrate that all hidden variable extensions of quantum mechanics are impossible.[3] This seemed to be the end of the matter. If hidden variables are impossible, why bother even to speculate about them?

And, indeed, silence prevailed for nearly twenty years. The dogmatic Copenhagen view prevailed, seeping into the mathematical formalism and becoming the quantum physicists' conscious or unconscious default interpretation. The physics community moved on and just got on with it, content to shut up and calculate.

Then David Bohm broke the silence.

In February 1951, Bohm published a textbook, simply called *Quantum Theory*, in which he followed the party line and dismissed the challenge posed by EPR's 'bolt from the blue', much as Bohr had done. But even as he was writing the book he was already having misgivings. He felt that something had gone seriously wrong.

Einstein welcomed the book, and invited Bohm to meet with him in Princeton sometime in the spring of 1951. The doubts over the interpretation of quantum theory that had begun to creep into Bohm's mind now crystallized into a sharply defined problem. 'This encounter had a strong effect on the direction of my research,' Bohm later wrote, 'because I then became seriously interested in whether a deterministic extension of quantum theory could be found.'[4] The Copenhagen interpretation had transformed what was really just a method of calculation into an *explanation* of reality, and Bohm was more committed to the preconceptions of causality and determinism than perhaps he had first thought.

In *Quantum Theory*, Bohm had asserted that 'no theory of mechanically determined hidden variables can lead to *all* of the results of the quantum theory.'[5] This was to prove to be a prescient statement. Bohm went on to develop a derivative of the EPR thought experiment which he published in a couple of papers in 1952 and which he elaborated in 1957 with Yakir Aharonov.[6] This is based on the idea of fragmenting a diatomic molecule (such as hydrogen, H_2) into two atoms.

Now, elementary particles are distinguished not only by their properties of electric charge and mass, but also by a further property which we call *spin*. This choice of name is a little unfortunate, and arises because some physicists in the 1920s suspected that an electron behaves rather like a little ball of charged matter, spinning around on its axis much like the Earth spins as it orbits the Sun. This is not what happens, but the name stuck.

The quantum phenomenon of spin is indeed associated with a particle's intrinsic *angular momentum*, the momentum we associate with rotational motion. Because it also carries electrical charge, a spinning electron behaves like a tiny magnet. But don't think this happens because the electron really is spinning around its axis. If we really wanted to push this analogy, then we would need to accept that an electron must spin *twice* around its axis to get back to where it started.* The electron has this property because it is a matter particle called a *fermion* (named for Enrico Fermi). It has a characteristic spin quantum number of ½ and two spin orientations—two directions the tiny electron magnet can 'point' in an external magnetic field. We call these 'spin up' (↑) and 'spin down' (↓). Sound familiar?

The chemical bond holding the atoms together in a diatomic molecule is formed by overlapping the 'orbits' of the electrons of the two atoms and by *pairing* them so that they have opposite spins—↑↓. In other words, the two electrons in the chemical bond are entangled. Bohm and Aharonov imagined an experiment in which the chemical bond is broken in a way that preserves the spin orientations of the electrons (actually, preserving the electrons' total angular momentum) in the two atoms. We would then have two atoms—call then atom A and atom B—entangled in spin states ↑ and ↓.

Bohm and Aharonov brought the EPR experiment down from the lofty heights of pure thought and into the practical world of the physics laboratory. In fact, the purpose of their 1957 paper was to claim that experiments capable of measuring correlations between distant entangled particles had already been carried

---

* Think about it like this. Make a Möbius band by taking a length of tape, twisting it once and joining the ends together so the band is continuous and seamless. What you have is a ring of tape with only one 'side' (it doesn't have distinct outside and inside surfaces). Now picture yourself walking along this band. You'll find that, to get back to where you start, you need to walk twice around the ring.

out. For those few physicists paying attention, Bohm's assertion and the notion of a practical test suggested some mind-blowing possibilities.

John Bell was paying attention. In 1964, he had an insight that was completely to transform questions about the representation of reality at the quantum level. After reviewing and dismissing von Neumann's 'impossibility proof' as flawed and irrelevant, he derived what was to become known as *Bell's inequality*. 'Probably I got that equation into my head and out on to paper within about one weekend,' he later explained. 'But in the previous weeks I had been thinking intensely all around these questions. And in the previous years it had been at the back of my head continually.'[7]

Recall from Chapter 4 that the EPR experiment is based on the creation of a pair of entangled particles, A and B, which we now assume to be atoms. Because the total angular momentum is conserved, we know that the atoms must possess opposite spin-up and spin-down states, which we will continue to write as $A_\uparrow B_\downarrow$ and $A_\downarrow B_\uparrow$. We assume that the atoms A and B separate as 'locally real' particles, meaning that they maintain separate and independent identities and quantum properties as they move apart.

We further assume that making any kind of measurement on A can in no way affect the properties and subsequent behaviour of B. Under these assumptions, when we measure A to be in an $\uparrow$ state, we then know *with certainty*, that B must be in a $\downarrow$ state. There is nothing in quantum mechanics that explains how this can happen, the theory is incomplete, and we have a big problem. This is the essence of EPR's original challenge.

But is any of this really so mysterious? Bell was constantly on the lookout for everyday examples of pairs of objects that are spatially separated but whose properties are correlated, as these provide accessible analogues for the EPR experiment. He found a perfect example in the dress sense of one of his colleagues at CERN, Reinhold Bertlmann. Some years later, Bell wrote:[8]

> The philosopher in the street, who has not suffered a course in quantum mechanics, is quite unimpressed by Einstein–Podolsky–Rosen correlations. He can point to many examples of similar correlations in everyday life. The case of Bertlmann's socks is often cited. Dr Bertlmann likes to wear two socks of different colours. Which colour he will have on a given foot on a given day is quite unpredictable. But when you see that the first sock is pink you can be already sure that the second sock will not be pink. Observation of the first, and experience of Bertlmann, gives immediate information about the second. There is no accounting for tastes, but apart from that there is no mystery here. And is not this EPR business just the same?

This situation was sketched by Bell himself, and is illustrated in Figure 12.

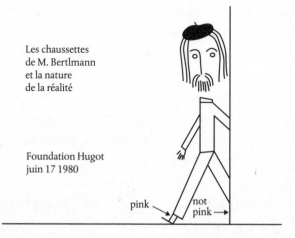

Figure 12 Bertlmann's socks and the nature of reality.

What if the quantum states of the atoms A and B are fixed by the operation of some local hidden variable at the moment they are formed and, just like Bertlmann's socks, the atoms move apart in already pre-determined quantum states? This seems to be perfectly logical, and entirely compatible with our first instincts. We can't deny an element of randomness, just as we can't deny that Bertlmann may choose at random to wear a pink sock on his left or right foot, so the hidden variable may randomly produce the result $A_\uparrow B_\downarrow$ or $A_\downarrow B_\uparrow$. But, so long as the spins of atoms A and B are always opposed, it would seem that all is well with the laws of physics.

How would this work? Well, we have no way of knowing what this hidden variable might be or what it might do, but here we are on the shores of Metaphysical Reality where we're perfectly at liberty to speculate. So, let's suppose that each atom has a further property we know nothing about, but which we presume acts to pre-determine the spins of A and B in any subsequent measurements. We could think of this property in terms of a tiny pointer, tucked away inside each atom. This can point in any direction in a sphere. When atoms A and B are formed by breaking the bond in the molecule, their respective pointers are firmly fixed in position, but they are constrained by the conservation of angular momentum always to be fixed in opposite directions.

The atoms move apart, the pointers remaining fixed in their positions. Atom A, over on the left, passes between the poles of a magnet, which allows us to measure its spin orientation.[*]

---

[*] This is known as a Stern–Gerlach apparatus, named for physicists Otto Stern and Walter Gerlach, who demonstrated the effect in 1922 using silver atoms. A beam of silver atoms passed between the poles of a magnet splits into two equal halves—one half is bent upwards towards the north pole, the other downwards towards the south pole—consistent with a random (50:50) alignment of the spins of the atoms' outermost electron, $\uparrow$ and $\downarrow$.

Atom B, on the right, passes between the poles of another magnet, which is aligned in the same direction as the one on the left. We'll keep this really simple. If the pointer for either A or B is projected anywhere onto the 'northern' semicircle, defined in relation to the north pole of its respective magnet (shown as the shaded area in Figure 13), then we measure the atom to be in an ↑ state. If the pointer lies in the 'southern' semicircle (the unshaded area), then we measure the atom to be in a ↓ state. Figure 13 shows how a specific (but randomly chosen) orientation of the pointers leads to the measurement outcome $A_\uparrow B_\downarrow$.

In a sequence of measurements on identically prepared pairs of atoms, we expect to get a random series of results: $A_\uparrow B_\downarrow$, $A_\uparrow B_\downarrow$, $A_\downarrow B_\uparrow$, $A_\uparrow B_\downarrow$, $A_\downarrow B_\uparrow$, $A_\downarrow B_\uparrow$, and so on. If we assume that in each pair the pointer projections can be randomly but uniformly distributed over the entire circle, then in a statistically significant number of measurements we can see that there's a 50% probability of getting a combined $A_\uparrow B_\downarrow$ result.

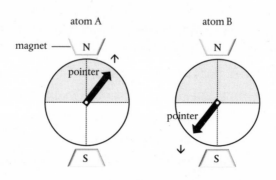

measurement outcome: $A_\uparrow B_\downarrow$

Figure 13 In a simple local hidden variable account of the correlation between the spins of entangled hydrogen atoms, we assume the measurement outcomes are predetermined by a 'pointer' in each atom whose direction is fixed at the moment the atoms are formed.

This is where Bell introduces a whole new level of deviousness. Figure 13 shows an experiment in which the magnets are aligned—the north poles of both magnets lie in the same direction. But what if we now *rotate* one of the magnets relative to the other? Remember that the 'northern' and 'southern' semicircles are defined by the orientation of the poles of the magnet. So if the magnet is rotated, so too are the semicircles. But, of course, we're assuming that the hidden variable pointers themselves are fixed in space at the moment the atoms are formed—the directions in which they point are supposedly determined by the atomic physics and can't be affected by how we might *choose* to orientate the magnets in the laboratory. The atoms are assumed to be locally real.

Suppose we conduct a sequence of three experiments:

	Orientation of magnet for Atom A	Orientation of magnet for Atom B	Difference in magnet angles
Experiment #1	0°	135°	135°
Experiment #2	135°	45°	90°
Experiment #3	0°	45°	45°

Figure 14 shows how rotating one magnet clockwise relative to the other affects the measurement outcomes for the same pointer directions used in Figure 13. In experiment #1, rotating the magnet for atom B by 135° means that the pointer for B now predetermines an ↑ state, giving the result $A_\uparrow B_\uparrow$. This doesn't mean that we've broken any conservation laws—the hidden variable pointers for A and B still point in opposite directions. It just means that we've opened up the experiment to a broader range of outcomes: rotating the magnet for atom B means that both $A_\uparrow B_\uparrow$ and $A_\downarrow B_\downarrow$ results have now become permissible. And, as

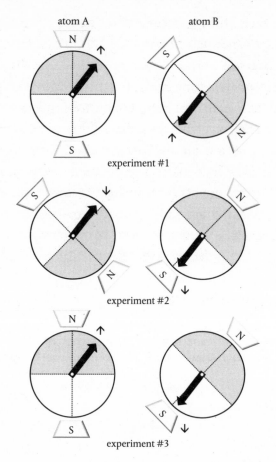

atom A   atom B

experiment #1

experiment #2

experiment #3

Figure 14 Bell introduced a whole new level of deviousness into the Bohm–Aharanov version of the Einstein–Podolsky–Rosen experiment by rotating the relative orientations of the magnets.

the probabilities for all the possible results must still sum to 100%, we can see that the probabilities for $A_\uparrow B_\downarrow$ and $A_\downarrow B_\uparrow$ must therefore fall.

The question I want to ask for each of these experiments is this: What is the probability of getting the outcome $A_\uparrow B_\downarrow$? Before rushing to find the answers, I'd like first to establish some numerical relationships between the probabilities for this outcome in each of the experiments. I don't want to get bogged down here in the maths, so I propose to do this pictorially.[9]

Imagine that we 'map out' the individual ↑ and ↓ results—irrespective of whether these relate to A or B—for each orientation of the magnets. For an orientation of 0°, we divide a square into equal upper and lower halves. For 135°, we divide the square into equal left and right halves. We have to be a bit more imaginative for the third 45° orientation, as we have only two dimensions to play with, so we draw a circle inside the square, such that the area of the circle is equal to the area that lies within the square, but outside the circle. This gives us

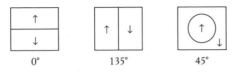

We can now combine these into a single map, which allows us to chart the $A_↑B_↓$ results for each of our experiments. For example, in experiment #1, results in which A is measured to be ↑ and B is measured to be ↓ occupy the top right-hand corner of the map, marked below in grey. Likewise for experiments #2 and #3:

experiment #1   experiment #2   experiment #3

We should note once again that this will only work if we can assume that atom A and atom B are entirely separate and distinct, and that making measurements on one can in no way affect the outcomes of measurements on the other. We must assume the atoms to be locally real.

If it helps, think of the grey areas in these diagrams as the places where we would put a tick every time we get an $A_↑B_↓$ result in each experiment. We carry out each experiment on exactly the same numbers of pairs of atoms, and we count up how many ticks we have. The number of ticks divided by the total number of pairs we studied then gives us the *probability* for getting the result $A_\uparrow B_\downarrow$ in each experiment.

In fact, these diagrams represent *sets* of numbers. So let's have some fun with them. We can write the set for experiment #1 as the sum of two smaller subsets:

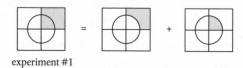

experiment #1

Likewise, we can write the set for experiment #2 as

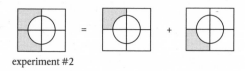

experiment #2

If we now add these two expressions together, we get

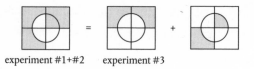

experiment #1+#2   experiment #3

We can't exclude the possibility that the last subset in this expression won't have some ticks in it, but I think you'll agree that it is perfectly safe for us to conclude that

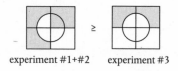

experiment #1+#2   experiment #3

where the symbol ≥ means 'is greater than or equal to'. This is Bell's inequality. It actually has nothing whatsoever to do with quantum mechanics or hidden variables. It is simply a logical conclusion derived from the relationships between independent sets of numbers. It is also quite general. It does not depend on what kind of hidden variable theory we might devise, provided this is locally real. This generality allowed Bell to formulate a 'no-go' theorem: 'If the [hidden variable] extension is local it will not agree with quantum mechanics, and if it agrees with quantum mechanics it will not be local.'[10] A complementary 'no-go' theorem was devised in 1967 by Simon Kochen and Ernst Specker.[11]

What this form of Bell's inequality says is that the probability of getting an $A_\uparrow B_\downarrow$ result in experiment #1, when added to the probability of getting an $A_\uparrow B_\downarrow$ result in experiment #2, must be greater than or at least equal to the probability of getting an $A_\uparrow B_\downarrow$ result in experiment #3.

Now, we can deduce these probabilities from our simple local hidden variable theory by examining the overlap between the 'northern' semicircles for the two magnets in each experiment, and dividing by 360°. We know from Figure 13 that complete overlap of 180° (the magnets are aligned) means a probability for $A_\uparrow B_\downarrow$ of 50%. In experiment #1, the overlap is reduced to 45°, and the probability of getting $A_\uparrow B_\downarrow$ falls to 12½%:

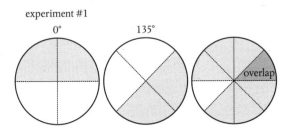

In experiment #2, the overlap is 90° (25%):

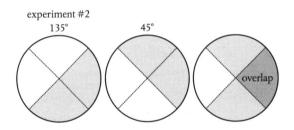

And in experiment #3 the overlap is 135° (37½%):

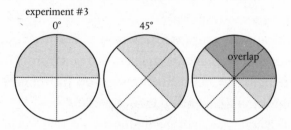

These are summarized on the left in the following:

	Probability of getting $A_\uparrow B_\downarrow$ local hidden variables	Probability of getting $A_\uparrow B_\downarrow$ quantum mechanics
Experiment #1	12½%	7.3%
Experiment #2	25%	25%
Experiment #3	37½%	42.7%

If we now add the probabilities for #1 and #2 together, then according to a local hidden variables theory we get a total of 37½%, equal to the probability for #3 and therefore entirely consistent with Bell's inequality.

So, what does quantum mechanics (without hidden variables) predict? I don't want to go into too much detail here. Trust me when I tell you that the quantum-mechanical prediction for the probability of getting an $A_\uparrow B_\downarrow$ result is given by half the square of the cosine of half the angle between the magnets. The quantum-mechanical predictions are summarized in the right-hand column of the table above. If we put these predictions into Bell's inequality, we get the result that 7.3% + 25% = 32.3% *must be greater than or equal to* 42.7%.

What we've discovered is that *quantum mechanics violates Bell's inequality.* It predicts that the extent of the correlation between atoms A and B can sometimes be greater, sometimes less, than *any* local hidden variable theory can allow.

This is such an important result that it's worth taking some time to recap, to ensure we understand how we got here. EPR sought to expose the incompleteness of quantum mechanics in a thought experiment involving a pair of entangled particles. If we adopt a realist interpretation of the wavefunction, and we assume that the particles are locally real and measurements on one can't in any way influence the outcomes of measurements on the other, then something is surely missing. Bohm and Aharonov adapted this experiment and showed how it might provide a practical test. Bell went further, introducing a whole new level of deviousness and devising Bell's inequality.

Here, indeed, is a direct test: quantum mechanics versus local hidden variables. Which is right? Is Bell's inequality violated *in practice*? This is more than enough reason to get back on board the Ship of Science and set sail for Empirical Reality.

Bell wrote his paper in 1964 but, due to a mix-up, it wasn't published until 1966.[12] It took about another ten years for experimental science to develop the degree of sophistication needed to begin to produce some definitive answers.

Although this kind of experimentation continues to this day, perhaps the most famous of these tests was reported in the early 1980s by Alain Aspect and his colleagues at the University of Paris. These were not based on entangled atoms and magnets. Instead they made use of pairs of entangled photons produced in a 'cascade' emission from excited calcium atoms.

Like electrons, photons also possess spin angular momentum, but there's a big difference. Photons are 'force particles'. They carry the electromagnetic force and are called bosons (named for Satyendra Nath Bose), and have a spin quantum number of 1. Because photons travel at the speed of light, there are only two spin orientations which we associate with left-circularly (☋) and right-circularly (☊) polarized light, as judged from the perspective of the *source* of the light. Now, the outermost electrons in a calcium atom sit in a spherical orbit with their spins paired and zero angular momentum. So, when one of these absorbs a photon and is excited to a higher-energy orbit, it picks up a quantum of angular momentum from the photon. This can't go into the electron's spin, since this is fixed. It goes instead into the electron's orbital motion, pushing it into an orbit with a different shape, from a sphere to a dumbbell—look back at Figure 6c.

But if we now hit the excited calcium atom with *another* photon, we can excite the electron left behind in the spherical orbit also into the dumbbell-shaped orbit. There are now three possible quantum states, depending on how the spin and orbital angular momenta of the electrons combine together. In one of these the angular momentum cancels to zero.

Although this state is very unstable, the calcium atom can't simply emit a photon and return to the lowest-energy spherical orbit. This would involve a transition with no change in angular momentum, and there's simply no photon for that. I suspect you can see where this is going. Instead, the atom emits *two* photons in rapid succession. One of the photons has a wavelength corresponding to green (we'll call this photon A) and the other is blue (photon B). As there can be no net change in angular momentum, and angular momentum must be conserved, the photons *must* be emitted with opposite states of circular polarization.

The photons are entangled.

The advantage of using photon polarization rather than the spin of electrons or atoms is that we can measure the polarization of light in the laboratory quite easily using polarizing analysers, such as calcite crystals.[13] We don't need to use unwieldy magnets.

One small issue. Polarizing analysers don't measure the circular polarization states of photons; they measure horizontal (↔) or vertical (↕) polarization.* But that's okay. A left- or right-circularly polarized photon incident on a linear polarizing analyser orientated vertically has a 50% probability of being transmitted. Likewise for an analyser orientated horizontally. And we know well enough by now that a total wavefunction expressed in a basis of left- and right-circular polarization states can be readily changed to a basis of horizontal and vertical polarization states.

Just like Bell's devious experiment with magnets, the analysers used to measure the polarization states of both photons A and B were mounted on platforms that could be rotated relative to one another. This experiment with entangled photons is entirely equivalent to Bell's.

One other important point. The detectors for each photon were placed 13 metres apart, on opposite sides of the laboratory. It would take about 40 billionths of a second for any kind of signal travelling at the speed of light to cross this distance. But the experiment was set up to detect pairs of photons A and B arriving within a window of just 20 billionths of a second. In other words, any spooky quantum influences passing between the photons—allowing measurements on one to affect the other—would need to travel faster than the speed of light.

We're now firmly on the shores of Empirical Reality, and we must acknowledge that the real world can be rather unco-operative. Polarizing analysers 'leak', so they don't provide 100% accuracy. Not all the photons emitted can be 'gathered' and channelled into their respective detectors, and the detectors themselves can be quite inefficient, recording only a fraction of

---

* Polaroid sunglasses reduce glare by filtering out horizontally polarized light.

the photons that are actually incident on them. Stray photons in the wrong place at the wrong time can lead to miscounting the number of pairs detected.

Some of these very practical deficiencies can be compensated by extending the experiment to a fourth arrangement of the analysers, and writing Bell's inequality slightly differently. For the particular set of arrangements that Aspect and his colleagues studied, Bell's inequality places a limit for local hidden variables of less than or equal to 2. Quantum mechanics predicts a maximum of 2 times the square root of 2, or 2.828. Aspect and his colleagues obtained the result 2.697, with an experimental error of ±0.015, a clear violation of Bell's inequality.[14]

These results are really quite shocking. They confirm that if we want to interpret the wavefunction realistically, the photons appear to remain mysteriously bound to one another, sharing a single wavefunction, until the moment a measurement is made on one or the other. At this moment the wavefunction collapses and the photons are 'localized' in polarization states that are correlated to an extent that simply cannot be accounted for in any theory based on local hidden variables. Measuring the polarization of photon A *does* seem to affect the result that will be obtained for photon B, and vice versa, even though the photons are so far apart that any communication between them would have to travel faster than the speed of light.

Of course, this was just the beginning. For those physicists with deeply held realist convictions, there just had to be something else going on. More questions were asked: What if the hidden variables are somehow influenced by the way the experiment is set up? This was just the first in a series of 'loopholes', invoked in attempts to argue that these results didn't necessarily rule out *all* the local hidden variable theories that could possibly be conceived.

Aspect himself had anticipated this first loophole, and performed further experiments to close it off. The experimental arrangement was modified to include devices which could randomly switch the paths of the photons, directing each of them towards analysers orientated at different angles. This prevented the photons from 'knowing' in advance along which path they would be travelling, and hence through which analyser they would eventually pass. This is equivalent to changing the relative orientations of the two analysers *while the photons were in flight*. It made no difference. Bell's inequality was still violated.[15]

The problem can't be made to go away simply by increasing the distance between the source of the entangled particles and the detectors. Experiments have been performed with detectors located in Bellevue and Bernex, two small Swiss villages outside Geneva almost 11 kilometers apart.[16] Subsequent experiments placed detectors in La Palma and Tenerife in the Canary Islands, 144 kilometers apart. Bell's inequality was still violated.[17]

Okay, but what if the hidden variables are still somehow sensitive even to random choices in the experimental setup, simply because these choices are made on the same timescale? In experiments reported in 2018, the settings were determined by the colours of photons detected from quasars, the active nuclei of distant galaxies. The random choice of settings was therefore already made nearly eight billion years *before* the experiment was performed, as this is how long it took for the trigger photons to reach the Earth. Bell's inequality was still violated.[18]

There are other loopholes, and these too have been closed off in experiments involving both entangled photons and ions (electrically charged atoms). Experiments involving entangled *triplets* of photons performed in 2000 ruled out all manner of locally realistic hidden variable theories without recourse to Bell's inequality.[19]

If we want to adopt a realistic interpretation, then it seems we must accept that this reality is *non-local* or, at the very least, it violates *local causality*.

But can we still meet reality halfway? In these experiments, we assume that the properties of the entangled particles are governed by some, possibly very complex, set of hidden variables. These possess unique values that predetermine the quantum states of the particles and their subsequent interactions with the measuring devices. We further assume that the particles are formed with a statistical distribution of these variables determined only by the physics and not by the way the experiment is set up.

Local hidden variable theories are characterized by two further assumptions. In the first, we assume (as did EPR) that the *outcome* of the measurement on particle A can in no way affect the outcome of the measurement on B, and vice versa. In the second, we assume that the *setting* of the device we use to make the measurement on A can in no way affect the outcome of the measurement on B, and vice versa.

The experimental violation of Bell's inequality shows that one or other (or both) of these assumptions is invalid. But, of course, these experiments don't tell us which.

In a paper published in 2003, Nobel laureate Anthony Leggett chose to drop the setting assumption. This admits that the behaviour of the particles and the outcomes of subsequent measurements *is* influenced by the way the measuring devices are set up. This is still all very spooky and highly counterintuitive:[20]

> nothing in our experience of physics indicates that the orientation of distant [measuring devices] is either more or less likely to affect the outcome of an experiment than, say, the position of the keys in the experimenter's pocket or the time shown by the clock on the wall.

By keeping the outcome assumption, we define a class of non-local hidden variable theories in which the individual particles possess defined properties before the act of measurement. What is actually measured will of course depend on the settings, and changing these settings will somehow affect the behaviour of distant particles (hence, 'non-local'). Leggett referred to this broad class of theories as 'crypto' non-local hidden variable theories. They represent a kind of halfway house between strictly local and completely non-local.

He went on to show that dropping the setting assumption is in itself still insufficient to reproduce all the results of quantum mechanics. Just as Bell had done in 1964, he derived an inequality that is valid for all classes of crypto non-local hidden variable theories but which is predicted to be violated by quantum mechanics. At stake then was the rather simple question of whether quantum particles have the properties we assign to them *before the act of measurement*. Put another way, here was an opportunity to test whether quantum particles have what we might want to consider as 'real' properties *before* they are measured.

The results of experiments designed to test Leggett's inequality were reported in 2007 and, once again, the answer is pretty unequivocal. For a specific arrangement of the settings in these experiments, Leggett's inequality demands a result which is less than or equal to 3.779. Quantum mechanics predicts 3.879, a violation of less than 3%. The experimental result was 3.8521, with an error of ±0.0227. Leggett's inequality was violated.[21] Several variations of experiments to test Leggett's inequality have been performed more recently. All confirm this general result.

It would seem that there is no grand conspiracy of nature that can be devised which will preserve locality. In 2011, Mathew Pusey, Jonathan Barrett, and Terry Rudolph at Imperial College in London published another 'no-go' theorem.[22] In essence, this says that any kind of hidden variable extension in which the wavefunction is interpreted purely statistically *cannot* reproduce all the predictions of quantum mechanics.

The 'PBR theorem', as it is called, sparked some confusion and a lot of debate when it was first published.[23] It was positioned as a theorem which rules out all manner of interpretations in which the wavefunction represents 'knowledge' in favour of interpretations in which the wavefunction is considered to be real. But 'knowledge' here is qualified as knowledge derived from the *statistics* of whatever it is that is assumed to underlie the physics and which is further assumed to be objectively real. Whilst it rules in favour of realist interpretations that are not based on statistics, it does not rule out the kinds of anti-realist interpretations which we considered in Chapters 5 and 6.

We should note in passing that whilst local or crypto non-local hidden variable theories have been all but ruled out by these experiments, they underline quite powerfully how realistic interpretations have provided compelling reasons for the experimentalists to roll up their sleeves and get involved. In this case, the search for theoretical insight and understanding, in the spirit of Proposition #4 (see the Appendix), has encouraged some truly wonderful experimental innovations. The relatively new scientific disciplines of quantum information, quantum computing, and quantum cryptography have derived in part from efforts to resolve these foundational questions and to explore the curious phenomenon of entanglement, even though the search for meaningful answers has so far proved fruitless.

But we must now confront the conclusion from all the experimental tests of Bell's and of Leggett's inequalities. In any realistic interpretation in which the wavefunction is assumed to represent the real physical state of a quantum system, *the wavefunction must be non-local or it must violate local causality*.

Okay, so let's see what that means.

8

Quantum Mechanics is Incomplete
SO WE NEED TO ADD
SOME OTHER THINGS

Pilot Waves, Quantum Potentials,
and Physical Collapse Mechanisms

Einstein was not alone in searching for ways to reintroduce causality and determinism in a realistic interpretation of quantum mechanics. De Broglie was looking, too, and at the fifth Solvay Conference in Brussels in 1927 he presented his own 'double solution' theory, involving both pilot waves and 'probability waves'. But if de Broglie had been hoping for support from Einstein at the conference, he was disappointed. Other than suggesting that de Broglie was searching in the right direction, Einstein remained impassive.

De Broglie's theory quickly evolved into a more familiar pilot wave theory, in which the wavefunction guides the paths of physically real particles. But further discussions (most notably with Pauli) raised doubts in his own mind about its validity and, by early 1928, he had all but abandoned it. He did not include it in a course on wave mechanics he taught at the Faculté des Sciences in Paris later that year. In fact, de Broglie became a convert to the Copenhagen orthodoxy.

Bohm's encounter with Einstein in 1951 encouraged him to look again and think more deeply. As I explained in Chapter 4, the Copenhagen interpretation is formally embedded in the standard quantum formalism through the structure of its axioms, and especially Axiom #1 (for a reminder, see the Appendix). Rather than accept this at face value, Bohm decided to explore the *possibility* that other descriptions and hence other interpretations are conceivable *in principle*.

He began by reworking Schrödinger's wave equation, assuming the existence of a real particle following a real path through space, its motion tied to the wave through the imposition of a 'guidance condition' which determines its velocity. What this means is that the motion of the particle is governed by the classical potential energy of the system—the steepness of the hill in my earlier Sisyphus analogy—*and* a second, so-called *quantum potential*. The latter is firmly non-classical and non-local and is alone responsible for the introduction of quantum effects in what would otherwise be an entirely classical description.

Soon afterwards, Bohm realized that he had rediscovered de Broglie's pilot wave theory, and the approach is now known variously as de Broglie–Bohm theory or Bohmian mechanics. It was Bell's interest in this theory that led him to devise his theorem and his inequality. He wanted to know whether, in any hidden variable interpretation, non-locality is inevitable (it is).

Note that in the de Broglie–Bohm interpretation, the wavefunction itself is very much part of the reality being described—it is not simply a convenient way of summarizing statistical behaviour derived from some kind of underlying, hidden reality. Consequently, it is not ruled out by the PBR no-go theorem. There is a statistical flavour, but this relates not to the wavefunction but to the (presumably random) spread of initial positions and velocities of the physically real particles that are guided by it.

These initial conditions determine which paths the particles will subsequently follow. A statistical distribution of initial conditions gives rise to a statistical distribution over the available paths.

Bohm's interest in the theory waned, but was resurrected in 1978 by the enthusiasm of Basil Hiley, his colleague at Birkbeck College in London, and the work of two young researchers, Chris Dewdney and Chris Philippidis. Seeing can sometimes be believing, and when Dewdney used de Broglie–Bohm theory to compute the hypothetical trajectories of electrons in a two-slit experiment (see Figure 15), the resulting picture provoked gasps of astonishment. In these simulations, each electron passes through one or other of the two slits, follows *one* of the predetermined paths, guided by the quantum potential, and is detected as a bright spot on the screen. As more electrons pass through the apparatus, the variation in their initial conditions means that

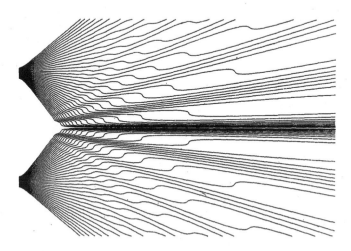

Figure 15 Particle trajectories in a two-slit experiment as predicted by de Broglie–Bohm theory.

they individually pass through different slits and follow different paths. The end result is a pattern of spots on the screen which reflects the grouping of the various paths—what we interpret as a two-slit interference pattern.

We can get some sense for how this might work from some recent, rather fascinating, experiments. John Bush and his colleagues at the Massachusetts Institute of Technology have reported on experiments in which a small oil droplet is bounced off the surface of a liquid.[1] Each bounce creates a set of ripples in the liquid which overlap and interfere, and the resulting interference pattern guides the subsequent motion of the droplet. As the ripples pass through two adjacent openings in a barrier they produce the familiar two-slit interference pattern (though there are now some doubts about this).[2] The 'walking' droplet then passes through one opening or the other, guided by the ripples to its destination. Although this is an entirely classical experiment, it is argued that these motions mimic the kinds of behaviours we might anticipate from de Broglie–Bohm theory, in which the oil droplet is replaced by a real quantum particle.

But such classical experiments cannot mimic the quantum potential which, true to its nature, exhibits some very peculiar behaviours. To explore these, let's return once more to our favourite quantum system consisting of particles prepared in a superposition of spin states ↑ and ↓. We pass these particles one at a time between the poles of a magnet. In de Broglie–Bohm theory, the quantum potential is split by the presence of the magnetic field into two equal but non-overlapping parts. One of these guides the particles upwards, and particles following this path will produce an ↑ result. The other guides the particles downwards, giving a ↓ result. The result we get depends on the initial conditions for each particle, but if a particle follows the upwards path, the quantum potential for the downward path doesn't disappear. Instead it persists as an 'empty wave'.

We have to assume that the particle following the upwards trajectory between the poles of the magnet interacts with some kind of detection device. Although this is likely to be of classical dimensions, we once again acknowledge that it is composed of quantum entities, and the first stages of the interaction between the particle and the detector will be quantum in nature. What follows is decoherence, but not in the sense of a convenient mathematical device designed to eliminate the interference terms from a pre-probability, as in the decoherent histories interpretation. In de Broglie–Bohm theory, the quantum potential is a real wave or field, and so decoherence is required to be a real physical mechanism. We'll examine the details later in this chapter.

In de Broglie–Bohm theory, the wavefunction doesn't physically collapse from all over space. The 'empty waves' persist, but these are no longer relevant to the measurement or the subsequent change in the state of our knowledge of the system. And the irreversible process of amplification from quantum to complex classical scales means that the measurement outcomes are decided long before we get to the scale of Schrödinger's cat.

This is all very fine, but once again such causal explanations come with a price tag, this one associated with the proliferation of empty waves. Of course, if empirical evidence for the existence of empty waves could be found, this would provide tremendous support for the theory. But, alas, we gain knowledge of the wavefunction only through measurements on the *particles* that are guided by it and, by definition, the empty waves are not associated with any particles—they're empty.

We're not quite done yet. Let's return again to the EPR experiment involving entangled atoms—A and B—formed in opposite spin states ↑ and ↓. Conservation of angular momentum demands that the spins must be aligned, such that if both magnets are orientated in the same direction (0°—see Figure 13), then the possible measurement outcomes are $A_\uparrow B_\downarrow$ and $A_\downarrow B_\uparrow$.

Suppose atom A moves off to the left and passes through the poles of a magnet orientated at 0°. We record an $\uparrow$ result. According to de Broglie–Bohm theory, this act of measurement performed on atom A instantaneously changes the quantum potential, and a kind of non-local quantum 'torque' is exerted on atom B. This action sets up the spin of atom B so that it will be detected in a $\downarrow$ state. Theorist Peter Holland writes that[3]

> The act of measurement on [A] polarizes [B] (in the direction of the analyzing field acting on [A]) and in any subsequent measurement on [B], the results will come out in the way predicted by quantum mechanics.

We see immediately that in this interpretation, non-locality really does imply action at a distance: 'the "spooky action-at-a-distance" is embraced as a fact of life from the outset. In this way any possibility of conflict with the empirical content of quantum mechanics is avoided.'[4]

We watch as the boulder of Sisyphus rolls down the hill. It reaches the bottom, where its motion is no longer influenced by the slope (the classical potential energy). If there's nothing in the way, and we can assume there's no friction to slow it down, then the boulder will trundle across the valley floor in a straight line at a constant speed according to Newton's first law of motion. But, although the valley is flat, in de Broglie–Bohm theory the quantum potential doesn't necessarily disappear here. The motion of the quantum boulder is still subject to some very spooky non-local influences, and Newton's first law no longer applies. Suppose its entangled partner passes through a measuring device positioned over in the next valley, causing ripples through the quantum potential. Imagine watching the first boulder rolling along the valley floor, when for no apparent reason it is hit by some invisible quantum torque which changes both its speed and direction.

De Broglie–Bohm theory restores causality and determinism. It eliminates the need to invoke a collapse of the wavefunction, but at the cost of accepting non-local spooky action at a distance. Make no mistake, this most definitely contradicts the spirit of Einstein's special theory of relativity, although advocates argue that the faster-than-light transfer of information implied by the theory cannot be extracted in any experiment that is also consistent with quantum mechanics. Particles A and B do indeed 'signal' each other over vast distances, by exerting effects on each other through the quantum potential, but we can do nothing useful with this. In this sense, de Broglie–Bohm theory and special relativity can peacefully—though rather uneasily—coexist.

Not surprisingly, Einstein was not enamoured of the approach Bohm had taken. In a letter to Born he wrote[5]

> Have you noticed that Bohm believes (as de Broglie did, by the way, 25 years ago) that he is able to interpret the quantum theory in deterministic terms? That way seems too cheap to me.

Today the de Broglie–Bohm theory retains a small but dedicated following within the communities of concerned physicists and philosophers, but it remains firmly outside the mainstream of quantum physics and features in few textbooks on the subject. It is in all respects equivalent to quantum mechanics and yet it allows a profoundly different interpretation of events occurring at the quantum level, one which is much more in tune with our more intuitive metaphysical preconceptions about the way reality ought to work.

It has been argued that the reason we teach the standard quantum mechanical formalism and not de Broglie–Bohm theory is one of historical contingency.[6] Who can say what might have happened if de Broglie had not been dissuaded in 1927, and the grip of the Copenhagen orthodoxy had been less firm? Could some kind of pilot wave theory have become the standard, default interpretation? This is a potentially disturbing argument for anyone with an idealistic view of how science progresses. Disturbing because the choice between equivalent, competing rival interpretations for one of the most important foundational theories of physics might have been driven simply by the *order in which things happened* rather than more compelling arguments based on notions of truth or explanatory power.

In 1982, Bell wrote[7]

> Why is the pilot wave picture ignored in text books? Should it not be taught, not as the only way, but as an antidote to the prevailing complacency? To show that vagueness, subjectivity, and indeterminism, are not forced on us by experimental facts, but by deliberate theoretical choice?

Frankly, I'm not so sure. Science through the ages has made a habit of stripping irrelevant or unnecessary metaphysical elements from theories that can be shown to work perfectly well without them. Examples include Ptolemy's epicycles, phlogiston, the ether, and caloric, the substance once thought to be responsible for the phenomenon of heat. Even if something like de Broglie–Bohm theory had become established as the preferred interpretation in 1927, I suspect it is really too cumbersome to have survived in this form. Those utilitarian physicists less concerned about causality and determinism, and less obsessed with interpretation and meaning, would have quickly dispensed with the theory's superfluous elements in the interests of more efficient calculation.

If we are inclined to agree with Einstein and abandon this approach as 'too cheap', then we must acknowledge that we're running out of options. This means abandoning *all* kinds of hidden variable theories, whether local, crypto non-local, or purely non-local. If we still want to continue to insist on a realist interpretation of the wavefunction, then we have to accept that we're now in a bit of a bind.

We have no choice but to attempt to pick away at *one* of the more stubborn conundrums of quantum mechanics and hope to at least resolve this by adding a further physical ingredient to the formalism. Even modest success might then provide some clues as to how we might resolve some of the others.

We look again at the formalism and try to identify some vulnerability, some point of attack. And once again, the rather obvious place to focus our attention is the process of quantum measurement, the collapse of the wavefunction, and the 'shifty split' this implies between the quantum and classical worlds. Bell perfectly encapsulated the uneasiness we feel in any realistic interpretation of quantum measurement in an article published in 1990:[8]

> What exactly qualifies some physical systems to play the role of 'measurer'? Was the wavefunction of the world waiting to jump for thousands of years until a single-celled living creature appeared? Or did it have to wait a little longer, for some better qualified system...with a PhD?

So let's look a little more closely at quantum measurement. Consider a quantum system consisting of a large number of identical particles (such as photons or electrons). Physicists call such a collection an *ensemble*. Think of the particles acting together 'in concert', all playing from the same score just like an ensemble of musicians.

We prepare the particles so that they're all represented by a single total wavefunction which we can write as a superposition of ↑ and ↓, as before. These particles are still in a single quantum state: it just happens that this is a state represented by a superposition. Such an ensemble is then said to be in a *pure state*. We pass the particles through a measuring device, and the total wavefunction collapses randomly into a sequence of measurement outcomes. After all the particles have passed through the device (and we assume they haven't been destroyed in the process) we would expect to have a 50:50 mixture of particles that are individually in the ↑ or ↓ state. *The measurement has transformed the ensemble from a pure state into a mixture.* Von Neumann showed that such a transformation is associated with an increase in the *entropy* of the system.

Entropy is a thermodynamic quantity that we tend to interpret rather crudely as the amount of 'disorder' in a system. For example, as a block of ice melts, it transforms into a more disordered, liquid form. As liquid water is heated to steam, it transforms into an even more disordered, gaseous form. The measured entropy of water increases as water transforms from solid to liquid to gas.

The second law of thermodynamics claims that, in a spontaneous change, entropy always increases. If we take a substance—such as air—contained in a closed system, prevented from exchanging energy with the outside world, then the entropy of the air will increase spontaneously and inexorably to a maximum as it reaches equilibrium with its surroundings. It seems intuitively obvious that the oxygen and nitrogen molecules and trace atoms of inert gas that make up the air will not all huddle together in one corner of the room in which I'm writing these words. Instead, the air spreads out to give a reasonably uniform pressure (thank goodness). This is the state with maximum entropy.

The second law ties entropy to the 'arrow of time', the experience that despite being able to move freely in three spatial dimensions—forward–back, left–right, up–down—we appear obliged to follow time in only one direction—forwards. Suppose we watch a video taken during a recent cocktail party. We watch as a smashed cocktail glass spontaneously reassembles itself from the wooden floor, refills with Singapore sling, and flies upwards through the air to return to a guest's fingers. We would quickly conclude that this reversal of the second law of thermodynamics signals that the video is playing *backwards* in time.

It would seem that quantum measurement, transforming a pure state into a mixture, is closely associated with entropy, and hence with the second law. From here it's a short step to the arrow of time and the notion of *irreversibility*. Just as that smashed cocktail glass ain't going to reassemble itself anytime soon, so we wouldn't expect a mixture of ↑ and ↓ quantum states to reassemble spontaneously into a superposition.

Bohr recognized the importance of the 'irreversible act' of measurement linking the quantum and classical worlds. Some years later, Wheeler wrote about an 'irreversible act of amplification'. The rather obvious truth is that we gain information about the quantum world only when we can amplify elementary quantum events and turn them into perceptible signals, such as the deflection of a gauge pointer. Perhaps a logical place to seek a *physical* collapse of the total wavefunction is right here, during this act. Schrödinger's cat is then spared the discomfort of being both dead and alive, because the irreversible act of amplification associated with registering a radioactive emission by the Geiger counter has already settled the matter, one way or the other.

The physicist Dieter Zeh was the first to note that the interaction of a wavefunction with a measuring apparatus and its 'environment' leads to rapid, irreversible decoupling of the components in a superposition in such a way that interference terms are suppressed. This is decoherence, which we first encountered in Chapter 6, but once again required to operate in a real physical sense, as we're wanting to apply this to a real wavefunction.

Let's stop and think about how this is supposed to work. As we've seen, the first step in a measurement interaction leads to the entanglement of whatever it is that is doing the interacting within the total wavefunction. This is followed by another interaction, then another, and another. We could suppose that this sequence continues until the classical measuring device is completely entangled within the total wavefunction, leading to the possibility of superpositions of, and interference between, classical-sized objects—such as a gauge pointing in two different directions at once, or a cat that is at once both alive and dead. Now, this would imply that the entanglement of the classical measuring device with the quantum system is entirely *coherent*.

But, of course, with some exceptions which I'll describe below, we *never* see superpositions of classical objects. The argument goes that as the sequence of interactions becomes more and more complex, the coherence required to maintain the integrity of the interference terms is quickly lost. The interference terms are essentially diluted away, and the wavefunction settles randomly on one outcome or the other. This is to all intents and purposes irreversible.

In this scenario the wavefunction doesn't 'collapse' instantaneously, as such. The time required for this kind of physical decoherence to take effect is obviously related to the size of the system under study and the number of particles in the device and the environment with which it interacts. The smaller the 'decoherence time', the faster the wavefunction loses coherence and evolves, together with the device and its environment, into what we recognize as a classical system.

Let's put this into some kind of perspective. A large molecule with a radius of about a millionth (10^{-6}) of a centimetre moving through the air is estimated to have a decoherence time on the order of a millionth of a trillionth of a trillionth (10^{-30}) of a second.[9] This means that the molecule becomes 'classical' within an unimaginably short time. If we remove the air and observe the molecule in a laboratory vacuum, we reduce the number of possible interactions and so we might increase the estimated decoherence time to 10^{-17} seconds, which is getting large enough to be imaginable (and potentially measurable). Placing the molecule in intergalactic space, where it is exposed only to the photons which constitute the cosmic microwave background radiation, increases the estimated decoherence time to 10^{12} seconds, meaning that the molecule may persist as a quantum system (for instance, in a superposition state) for a little under 32,000 years.

In contrast, a dust particle with a radius of a thousandth of a centimetre—a thousand times larger than the molecule—has a decoherence time of a microsecond (10^{-6} seconds), even in intergalactic space. The dust particle behaves classically even here, where the likely number of interactions with the environment is at its very lowest.

Clearly, for quantum entities such as photons, electrons, and individual atoms, the decoherence times will be longer in all the different environments considered. But when large numbers of interacting particles are involved, as in the interaction of the quantum systems with a classical measuring device and its environment, the decoherence time becomes extremely short and, for all practical purposes, the transition from quantum to classical behaviour can be assumed to be essentially instantaneous, and irreversible.

The kinds of timescales over which decoherence is expected to occur suggest that it will be very tricky—though perhaps not impossible—to catch a quantum system in the act of losing coherence in any conventional experiment. Recall, however, that this is one of those times where size really *does* matter. It is possible to find systems of intermediate scale, *between* quantum and classical, with decoherence times measured in microseconds to milliseconds. Decoherence has been observed directly in systems involving trapped beryllium ions, and highly excited rubidium atoms sitting inside cavities filled with microwave photons.[10] If decoherence isn't allowed to proceed too far, it is possible to reverse the process and watch as the initial wavefunction is recovered through *recoherence*.[11]

Diffraction and interference effects have been demonstrated at these intermediate scales using closed-cage molecules consisting of 60 carbon atoms (called buckminsterfullerene) and 70 carbon atoms (fullerene-70), and more recently in large organic molecules with up to 430 atoms.[12] In experiments performed on so-called superconducting quantum interference devices (SQUIDs), interference has been observed between states involving about a billion pairs of electrons *travelling in opposite directions* around a superconducting ring large enough to be visible to the naked eye.[13] Such systems have been described as the laboratory equivalents of Schrödinger cat states.

These experiments demonstrate that what we've so far thought of as the collapse of the wavefunction is not some philosophical or mathematical abstraction, but a real physical process that can be observed and quantified. They also demonstrate the lengths we have to go to in order to avoid exposure to an environment likely to induce rapid decoherence.

We might then wonder why, if quantum coherence is really so fragile and difficult to maintain, how is it possible for us routinely to observe interference effects requiring coherent superpositions of many photons? The answer is that the interactions occurring in an electromagnetic field involving large numbers of photons are primarily photon–photon interactions. Such interactions do happen but they are extremely weak. To a first approximation, photons do not interact with themselves at all and so do not represent a significant source of decoherence in an intense electromagnetic field. Coherence survives, and interference at large scales can be readily observed.

This is beginning to feel like real progress. A physical explanation for the collapse of the wavefunction must surely solve the problem of quantum measurement. And, if the collapse is a real physical thing, then this must mean just as surely that the wavefunction itself must be real. How could it not be? There has to be something physical to decohere.

But it should be evident by now that in quantum mechanics it pays never to get too carried away. Alas, decoherence does *not* solve the measurement problem, as Bell argued:[14]

> The idea that elimination of coherence, in one way or another, implies the replacement of 'and' by 'or', is a very common one among solvers of the 'measurement problem'. It has always puzzled me.

Decoherence suppresses the interference terms by diluting them over the vast number of states in the measuring device and its environment. We can argue that the evolving measurement forces a 'privileged basis', one that necessarily accords with the way the measurement is set up and hence with our classical experience. We make measurements and record that a particle was either ↑ or ↓, because this is what the device is set up to measure and so these outcomes are 'robust' to the effects of decoherence. But as a physical mechanism, decoherence can't *force* the choice between all the different measurement possibilities. That we randomly get either ↑ or ↓ remains essentially mysterious. Decoherence can't account for quantum probability; it can't convert ↑ *and* ↓ into ↑ *or* ↓.

The mathematical physicist Roger Penrose makes similar observations:[15]

> [Decoherence] does not help us to determine that the cat is actually either alive or dead... we need more... What we do not have is a thing which I call **OR** standing for *Objective Reduction*. It is an objective thing —either one thing or the other happens objectively. It is a missing theory. **OR** is a nice acronym because it also stands for 'or', and that is indeed what happens, one **OR** the other.

The theoretician Roland Omnès has called this the problem of 'objectification'. In the context of decoherence, it remains unsolved.

Okay, so that's a bit disappointing. But what about the implications of the experimental observation of decoherence for the reality of the wavefunction? Can we at least gain some solace from this?

Alas, no. Decoherence is a critically important mechanism, but it is not in itself a distinct interpretation of quantum mechanics. In fact, as we have already seen in the case of the consistent or decoherent histories interpretation, it is a mechanism that is most usefully employed *within* an interpretation to nail down the 'shifty split' between the quantum and classical domains. As such, it is equally at home in both realist and anti-realist interpretations.

Let me explain. We are sorely tempted to conclude that the physical process of decoherence implies a physically real quantum state. This is understandable. But we are still perfectly at liberty to suppose that decoherence applies to a system that we choose to represent in terms of a wavefunction that simply holds *information* about the quantum system. Instead of a real physical wavefunction, together with its interference terms, becoming diluted through the sequence of ever more complex interactions associated with a measurement, we follow the evolution of the *information* we presume it contains. As before, the forced choice between measurement possibilities is then just an unproblematic updating of our knowledge.

You might think that information as a concept seems too artificial or abstract to have any real physical consequences, but the relational and information-theoretic interpretations both deal with real physics. It's just that they don't require a realistic interpretation of the wavefunction to do so. And, what's more, we can just as easily base our arguments about the relation between quantum measurement and entropy on the information content of the wavefunction.

In 1948, the mathematician and engineer Claude Shannon developed an early but very powerful form of information theory. Shannon worked at Bell Laboratories in New Jersey, the prestigious research establishment of American Telephone and Telegraph (AT&T) and Western Electric (it is now the research and development subsidiary of Alcatel-Lucent). Shannon was interested in the efficiency of information transfer via communications channels such as telegraphy. He found that 'information' as a concept can have what we would normally regard as physical properties. Most notably, information has entropy, which we know today as 'Shannon entropy'. It is in many ways equivalent to the 'Von Neumann entropy' derived from quantum mechanics.

In 1961, this kind of logic led IBM physicist Rolf Landauer to declare that 'information is physical'. He was particularly interested in the processing of information in a computer. He concluded that when information is erased during a computation it is actually dumped into the environment surrounding the processor, adding to the entropy. This increase in entropy results in an increase in temperature: the environment surrounding the processor heats up. Anyone who has ever run a complex computation on their laptop computer will have noticed how, after a short while, the computer starts to get uncomfortably hot.

Landauer's famous statement requires some careful interpretation, but it's enough for now to note the direct connection between the processing of information and physical quantities such as entropy and temperature. It seems that 'information' is not an abstract concept invented by the human mind. It can have real, physical consequences.

The bottom line is that observation of physical decoherence can't be taken as evidence that the wavefunction represents the physically real state of a quantum system. Yes, there has to be something physical to decohere, but interpretations based on information will work just as well.

Decoherence provides a physical basis for understanding the transition between the quantum and the classical worlds, but it does so by relying on conventional quantum mechanics extrapolated to complex, large-scale systems. In this sense, the only thing 'added' to the formalism is a complexity that is otherwise absent or just ignored. Given that there is now some well-established experimental evidence for decoherence, it might appear strange that the phenomenon is often overlooked in student textbooks.[16] The simple truth is that decoherence does not remove the need for an interpretational framework and, once again, those physicists less concerned with interpretation and meaning tend not to fuss about the quantum-to-classical transition, because they really don't need to.

But there's nothing to stop us going a little further. Whilst enjoying this short visit to the shores of Metaphysical Reality, why not supplement quantum mechanics with a completely new physical mechanism, one that avoids the 'shifty split' without running into the problem of objectification? This is what physicists Giancarlo Ghirardi, Alberto Rimini, and Tullio Weber did in 1986. Their initial theory has subsequently been refined and extended by Philip Pearle and others, but for simplicity I will continue to refer to it here as GRW theory.[17]

GRW chose to add a new mathematical term to the quantum formalism. The new term has the effect of subjecting the wavefunction (interpreted realistically) to random, spontaneous localizations, or jumps, or 'hits', if you prefer. Instead of being left alone, to glide gracefully through space on a course set by the Schrödinger equation according to Axiom #5, the wavefunction is poked every now and then with the quantum equivalent of an electric prod, which forces it to collapse and curl up on itself rather like a startled hedgehog. To get this to work properly, GRW found that they needed to introduce two new physical constants. The first of these refers to the 'localization accuracy', which determines the dimensions of the function that combines with the total wavefunction describing the superposition when it is 'hit'. They fixed on a localization accuracy of about 10^{-5} centimetres.

The second of these new constants represents the mean frequency of spontaneous localizations. GRW set this to a value of 10^{-16} per second. This implies that the wavefunction is localized on average about once every few hundred million years. However, just like decoherence, this frequency is sensitively dependent on the *number* of interacting particles involved, such that a complex system localizes at an average frequency given by the number of particles times 10^{-16} per second.

This means that the wavefunction of a quantum system consisting of individual or small numbers of particles *never* localizes: it continues to evolve in time according to the Schrödinger equation. With these choices for the constants, there is no practical difference between the GRW theory and conventional quantum mechanics, at least for quantum systems. However, any kind of classical measuring device will consist of trillions and trillions of particles, and the mean localization frequency increases such that the wavefunction is localized (collapsed) within a few billionths of a second: '[Schrödinger's] cat is not both dead and alive for more than a split second.'[18]

There are obvious parallels with the mechanics of decoherence, but the GRW theory has the added advantage of forcing the selection of a specific measurement outcome, though this is still very much a random process associated with the spontaneous localizations.

There are versions of spontaneous collapse mechanisms that are variations on the GRW theme but which address issues of application, for example to quantum systems consisting of ensembles of identical particles, and which make the collapse mechanism continuous. We should accept that such mechanisms are bolted on to conventional quantum mechanics for no other reason than to satisfy our metaphysical preconceptions concerning the reality of the wavefunction (Proposition #3). But, although we might baulk at the rather ad hoc nature of this particular 'work around', we should also acknowledge that, once again, this realist extension has provoked sufficient curiosity to prompt a search for empirical evidence. In other words, GRW theory is an 'active' interpretation according to Proposition #4 (see the Appendix). Much of the experimental effort to detect decoherence in systems of intermediate scale serves as crucibles for potential evidence for spontaneous collapse mechanisms, too. Although to date no favourable evidence has yet been gathered, Ghirardi is optimistic: 'fully discriminating tests seem not to be completely out of reach'.[19] There's more on this in the remaining part of this chapter.

But introducing new physical constants is always less satisfactory than having the solution to the problem emerge naturally from the theory itself.

Recall that in later life Einstein tended to assume that all these problems would be resolved in an elusive grand unified theory. Quantum mechanics is clearly not the end. It is not finished. It relies on the assumption of a background space and time not much different from Newton's metaphysical absolutes, in contrast with Einstein's general theory of relativity, in which space and time are *emergent*. This poses a vitally important question. Can a quantum theory of gravity save us?

In general relativity, the action at a distance implied by the classical Newtonian force of gravity is replaced by a curved spacetime. The amount of curvature in a particular region of spacetime is related to the density of mass–energy present. Wheeler explained it this way: 'Spacetime tells matter how to move; matter tells spacetime how to curve.'[20]

This logic led a number of physicists, starting with Feynman, to suggest that the structure of spacetime itself may have a role to play in quantum mechanics. Lajos Diósi (in 1987) and later Roger Penrose (in 1996) proposed another kind of spontaneous collapse mechanism, now commonly referred to as Diósi–Penrose theory.[21] They argue that a superposition will begin to break down and eventually collapse into specific quantum states when it encounters a region of significant spacetime curvature. Unlike decoherence or the GRW theory, in which the *number* of particles is the key to the collapse, in Diósi–Penrose theory it is the *density* of mass–energy which is important, as this determines the extent of spacetime curvature around it. In his popular book *The Emperor's New Mind*, Penrose wrote[22]

My own point of view is that as soon as a 'significant' amount of space-time curvature is introduced, the rules of quantum linear superposition must fail. It is here that the complex-amplitude superpositions of potentially alternative states become replaced by probability-weighted actual alternatives—and one of the alternatives indeed *actually* takes place.

This is potentially quite neat. On a quantum scale, gravitational (spacetime curvature) effects are insignificant, leaving the wavefunction free to evolve according to the Schrödinger equation. But these effects become much more significant when the wavefunction encounters a classical measuring device. Of course, the quantum system is created in a laboratory which sits in Earth's gravitational field, so Penrose suggests that it is the *difference* in spacetime curvature in the two situations which triggers the collapse.

Diósi–Penrose theory explains how gravitationally induced decoherence or GRW-like spontaneous collapse mechanisms would work. But these arguments do not derive from a fully fledged quantum theory of gravity. The disappointment is that the current prime candidates for such a theory—loop quantum gravity (my personal favourite) and superstring theory—can't yet offer us any further clarity. There may indeed be insights to be gained, such as Smolin's suggestion that the non-local connections *between* quanta of space predicted by loop quantum gravity might explain non-locality in quantum mechanics.[23] But it's still too soon to be definitive about any of this. As far as I can tell, these theories of quantum gravity are still heavily dependent on quantum mechanics as a foundational theory, and in themselves do not require this particular foundation to be different than it is.*

It seems we're unlikely to get any answers directly from theory anytime soon. Despite this, we see yet again how proposals based

---

* Which is why both Rovelli and Smolin—who have long collaborated on loop quantum gravity—have in their own individual ways sought to come to terms with the interpretation of quantum mechanics.

on realistic interpretations serve to motivate the community of experimentalists. Any suggestion—sometimes no matter how tenuous—that reality might actually be different than we understand it to be in ways that can be probed in experiments is more than enough to pique their interest. It's more than enough reason to climb back aboard the Ship of Science and sail to the shores of Empirical Reality.

In his more recent book *Fashion, Faith and Fantasy*, Penrose summarizes 'various currently active proposals' to test these ideas.[24] One such proposal, involving macroscopic quantum resonators (MAQRO) installed in an orbiting satellite, aims to test the predictions of quantum mechanics for superpositions of objects with more than a hundred million atoms.[25] In these circumstances, it is hoped that the subtle differences between the predictions of quantum mechanics and the GRW and Diósi–Penrose theories might start to become accessible to experiment.

As space missions go, MAQRO is 'medium-sized'. It was first proposed to the European Space Agency (ESA) in 2010, and this proposal was substantially updated and resubmitted in 2015.[26] In September 2016 it was resubmitted in response to the ESA's call for 'New Science Ideas', and was selected for further detailed investigation by the ESA's Concurrent Design Facility during 2018.*

If the mission gets the go-ahead, it will build on all the knowledge from the recent highly successful LISA Pathfinder, launched in December 2015, which was designed to test technology capable of detecting gravitational waves in space.†

---

* For more details and to follow the project's progress, see http://maqro-mission.org/.
† LISA stands for Laser Interferometer Space Antenna.

The cost of the LISA Pathfinder mission was estimated to be €400 million back in 2011.* Given that, if it is commissioned, the MAQRO mission is unlikely to launch for another ten years or more, its budget requirements will inevitably be substantially larger than this. As I've said before, the metaphysical preconceptions that give rise to realist convictions among quantum physicists tend to come with a hefty price-tag.

For sure, the MAQRO mission will do more than test the representation of quantum reality in macroscopic superpositions, but the price to be paid refers to more than just a few philosophical conundrums (and the occasional headache).

* https://spacenews.com/lisa-pathfinder-proceed-despite-100-cost-growth/

9

Quantum Mechanics is Incomplete
BECAUSE WE NEED TO INCLUDE MY MIND (OR SHOULD THAT BE YOUR MIND?)

Von Neumann's Ego, Wigner's Friend, the Participatory Universe, and the Quantum Ghost in the Machine

Of course, the problem of the collapse of the wavefunction didn't originate in 1932 with the publication of von Neumann's *Mathematical Foundations of Quantum Mechanics*. But it's fair to say that his approach to quantum measurement really dragged the problem into the open, from where it has proceeded to torture the intellects of quantum physicists and philosophers for the past 90 years or so.

As we've seen, the anti-realists dismiss the collapse as a non-problem, no more difficult to understand than the abrupt change in our knowledge when we gain some new information. A few theorists more inclined to realist preconceptions have sought to identify physical mechanisms that act on physically real wavefunctions, designed in attempts to explain how *and* becomes *or*.

But what did von Neumann himself think was going on?

In his classic text, von Neumann clearly distinguished between two fundamentally different types of quantum process. The first, which he referred to as process **1**, is the discontinuous, irreversible transformation of a pure quantum state into a mixture, involving the 'projection' of some initial wavefunction into one of a set of possible measurement outcomes, with an accompanying increase in entropy. We now call this the collapse of the wavefunction, although von Neumann himself didn't use this terminology.* Process **2** is the continuous, deterministic, and completely reversible evolution of a wavefunction, governed by the Schrödinger equation according to Axiom #5 (see Appendix). These two processes are distinct: process **1** cannot happen in process **2**, and vice versa.

He then looked at quantum measurement from the perspective of three fundamental components, which he labelled **I, II,** and **III:**

I   is the quantum system under investigation;
II   is the physical measurement; and
III is the 'observer'.

He proceeded to demonstrate that if a quantum system **I** is present in a superposition of the measurement outcomes (for example, particle A is in a superposition of ↑ and ↓ states), then this will evolve smoothly and continuously according to process **2**. On encountering the measuring device, the wavefunction becomes entangled, but von Neumann saw no reason to suppose that quantum mechanics ceases to apply at this classical scale. The entangled wavefunction must then continue to evolve smoothly according to the Schrödinger equation. Process **2** still applies.

---

* He tended to avoid the word 'wavefunction', presumably as this is closely associated with Schrödinger's wave mechanics. He preferred to think of the description of quantum systems in terms of rather more abstract 'state functions' in a mathematical 'Hilbert space'. And, with some modifications, this is the description commonly taught to students today.

If we further entangle the device with a gauge, we know well by now that this gives rise to another superposition consisting of components $A_\uparrow$ and $A_\downarrow$.

Von Neumann could find no reason, based purely on the mathematics, to suppose that process 1 would have any role to play in the composite system I plus II. Process 2 applies equally to classical measuring devices and gauges as it does to quantum systems.

Schrödinger wouldn't publish the paper containing the reference to his famous cat for another three years, and von Neumann was already aware of the implications for an infinite regress. But his resolution of the problem was quite straightforward. If the quantum mechanics described by process 2 applies equally well to classical measuring devices, then there is again no good reason to suppose that it ceases to apply when considering the function of human sense organs, their connections to the brain, and the brain itself. Suppose that the laboratory has an overhead light which illuminates the screen of the gauge and some of the reflected light is gathered and focused at the observer's retinas. This triggers electrical signals in the observer's optic nerves, which travel to the visual cortex located at the back of the observer's brain.

We can choose simply to expand the definition of the 'quantum system' in I to include the particle A, the classical measuring device, the gauge, and the reflected light. Component II—the 'physical measurement'—then includes the observer's sensory apparatus and brain. The result is yet another superposition:

This implies that the observer enters into a superposition of 'brain states', and .

Von Neumann wrote:[1]

> Now quantum mechanics describes the events which occur
> in the observed portions of the world, so long as they do not
> interact with the observing portion, with the aid of process 2...,
> but as soon as such an interaction occurs, i.e. a measurement,
> it requires the application of process 1.

So what, then, did he have in mind with regard to the 'observing
portion' of the world? Based on conversations he had had with
his Hungarian compatriot Leo Szilard, von Neumann suggested
that component **III** consists of the observer's 'abstract ego'. In other
words, process 1—the collapse of the wavefunction—only occurs
when the measurement outcome is registered in the observer's
conscious mind.

The logic is pretty unassailable. No observer has ever reported
experiencing a superposition of brain states (or, at least, anyone
declaring that they have directly experienced such a superposition wouldn't be taken very seriously). Components **I** and **II** are
entirely 'mechanical' in nature—they involve physics and bio-
chemistry. We're left to conclude that because **III** is not mechanical, then this must be the place where the continuous evolution
of the wavefunction—process 2—breaks down, to be replaced
by process 1.

This conclusion is nevertheless quite extraordinary considering von Neumann's mission in the *Mathematical Foundations*,
which was to provide a much more secure mathematical basis
for quantum mechanics using Hilbert's axiomatic approach.
As I've already mentioned, these axioms (especially Axiom #1)
served to entrench the prevailing Copenhagen interpretation
directly in the formalism itself. Although Bohr was more
ambiguous about what this actually meant, Heisenberg's perspective was firmly anti-realist. Yet in his theory of quantum

measurement, von Neumann went substantially beyond the Copenhagen interpretation, ignoring Bohr's insistence on an arbitrary boundary between the quantum and classical worlds.

I would argue that the introduction of a role for consciousness in the measurement process represents an *addition* to the conventional quantum formalism, seemingly at odds with the 'nothing to see here' axiom. I guess von Neumann would have responded that Axiom #1 refers only to the *mathematical* structure, and the proposed addition of component **III** is decidedly non-mathematical. He wrote that: '**III** remains outside of the calculation'.[2]

As it stands, von Neumann's theory can still be interpreted in two fundamentally different ways. An anti-realist would agree with the description of components **I** and **II** and interpret **III** not as a physical collapse, but as the registering of the measurement outcome and the updating of the observer's state of knowledge about it. This most definitely involves the observer's conscious mind, but only in a passive sense, and takes us back to relational quantum mechanics, or information-theoretic interpretations, or QBism (take your pick).

But it seems that von Neumann held a different view. Component **III** is intended as the place where process **1** occurs, considered as *a real physical collapse*. His long conversations with Szilard concerned the latter's work on entropy reduction in thermodynamic systems through interference by intelligent beings, a variation on Maxwell's Demon.[3] The philosopher Max Jammer notes that this kind of paper 'marked the beginning of certain thought-provoking speculations about the effect of a *physical intervention* of mind on matter'.[4]

A real physical collapse implies a real wavefunction, and therefore a much more active role for the observer's conscious mind. It is for this reason that I include consciousness-causes-collapse theories in my collection of realist interpretations of quantum mechanics. That von Neumann wasn't specific on how he thought the wavefunction itself should be interpreted just adds to the confusion (but is pretty much par for the course in this business).

But now we need to ask ourselves: Just *who* is the observer? Let's return to the scenario we've considered a few times already, in which Alice makes a measurement in the laboratory but Bob is delayed in the corridor. I'm going to make one small adjustment. I'm going to replace Bob with renowned theorist Eugene Wigner. Alice and Wigner are close friends.[5]

We recall that Alice performs a measurement on a quantum system consisting of an ensemble of A particles prepared in a superposition of ↑ and ↓ states. Instead of a gauge the measuring device is now connected to a simple light switch. If the device records an ↑ result, the switch is not thrown and the light doesn't flash (⬛). If the device records a ↓ result, the switch is thrown and the light flashes (☀). She runs the experiment once, and observes the light flash.

Wigner is still in the corridor. As far as he is concerned, the total wavefunction that Alice just experimented on has the form of another superposition involving the measurement outcomes $A_↑$ and $A_↓$, the two possible states of the light, ⬛ and ☀, and Alice's possible brain states, 😐 and 😐:

Wigner now enters the laboratory. The following conversation ensues.

'Did you see the light flash?' asks Wigner.

'Yes,' replies Alice.

As far as Wigner is concerned, the measurement outcome has just registered in his conscious mind and the wavefunction collapses into the state described by $A_↓$☀😐.

But, after some reflection, he decides to probe his friend a little further.

'What did you feel about the flash before I asked you?'

Understandably, Alice is starting to get a little irritated. 'I told you already, I *did* see a flash,' she replies, testily.

Not wishing to put any further strain on his relationship with Alice, he decides to accept what she's telling him. He concludes that the wavefunction must have already collapsed into the state $A_\downarrow$ 🎇🧠 *before* he entered the laboratory and asked the question, and the above superposition that he took to be the correct description is, in fact, wrong. This superposition 'appears absurd because it implies that my friend was in a state of suspended animation before [she] answered my question'.[6] He wrote:

> It follows that the being with a consciousness must have a different role in quantum mechanics than the inanimate measuring device…It is not necessary to see a contradiction here from the point of view of orthodox quantum mechanics, and there is none if we believe that the alternative is meaningless, whether my friend's consciousness contains either the impression of having seen a flash or of not having seen a flash. However, to deny the existence of the consciousness of a friend to this extent is surely an unnatural attitude, approaching solipsism, and few people, in their hearts, will go along with it.

This is the *paradox of Wigner's friend*. To resolve it we must presume that the irreversible collapse of the wavefunction is triggered by the *first* conscious mind it encounters.

There's more. Nowhere in the physical world is it possible physically to act on an object without some kind of reaction. This is Newton's third law of motion. Should consciousness be any different? Although small, the action of a conscious mind in collapsing the wavefunction produces an immediate reaction— knowledge of the state of a system is irreversibly (and indelibly) generated in the mind of the observer. This reaction may lead to other physical effects, such as writing the result in a laboratory notebook, or the publication of a research paper, or the winning of a Nobel Prize. In this hypothesis, the influence of matter over mind is balanced by an influence of mind over matter.

If we introduce a role for consciousness in our representation of quantum mechanics, then we must acknowledge the truth of one of Wheeler's favourite phrases. He argued, paraphrasing Bohr, that 'No elementary phenomenon is a phenomenon until it is a registered (observed) phenomenon.'[7] Rather than interpret this registration process simply as an irreversible change in our knowledge of the system, Wheeler explored a more realistic interpretation in which the process is actually an irreversible *act of creation*. 'We are inescapably involved in bringing about that which appears to be happening.'[8]

Wheeler took some pains to separate the notion of a 'quantum phenomenon' from consciousness, arguing that it is the physical, irreversible act of amplification that brings the phenomenon about. But his phrase includes the word 'observed', and if we accept a role for conscious observation in quantum physics, then we arrive at more or less the same conclusions. If consciousness is required to collapse the wavefunction and 'make it real', then arguably the quantum state that it represents does not exist until it becomes part of the observer's conscious experience. It's a relatively small step from this to the rejection of Proposition #1. *Nothing* exists unless and until it is consciously experienced.

And, indeed, in 1977 Wheeler himself elaborated what was to become known as the 'participatory anthropic principle':[9]

Nothing speaks more strongly for this thesis than...the anthropic principle of [Brandon] Carter and [Robert] Dicke and...the indispensable place of the participating observer—as evidenced in quantum mechanics—in defining any useful concept of reality. No way is evident to bring these considerations together into a larger unity except through the thesis of 'genesis through observership'.

Here anthropic means 'pertaining to mankind or humans'. Although Wheeler would subsequently declare that the 'eye' of the observer 'could as well be a piece of mica',[10] it's virtually impossible to read this 1977 essay without concluding that this is about 'us', participants of a universe that we create by observing it. In their comprehensive review of anthropic reasoning, called *The Anthropic Cosmological Principle*, John Barrow and Frank Tipler accepted Wheeler's sentiments to be consistent with a version of what they call the strong anthropic principle: '*Observers are necessary to bring the universe into being.*'[11]

Okay. So on this particular visit to the shores of Metaphysical Reality, we've settled ourselves comfortably in a deckchair on the beach, with the Sun shining, shades on, sipping a margarita. We're here to stay a while. Whilst the logic is clear, what we're trying to do here is resolve two very deeply rooted philosophical conundrums—the collapse of the wavefunction and the nature of consciousness—simply by bringing them together. I have to say that this has never struck me as a particularly productive way to go.

By introducing consciousness into the mix, we invite an awful lot of further difficult questions. What is consciousness and how does it work? What does it mean when we say that consciousness is something other than 'mechanistic' and what evidence do we have for this? How is the collapse of the wavefunction supposed to be triggered by consciousness? Is the collapse of the wavefunction actually *responsible* for consciousness? Is the mind a quantum computer?

These questions are no doubt thought-provoking, but don't expect to find too many ready answers. The study of consciousness is the only discipline I've come across that is structured principally in terms of its *problems*. We have the 'hard problem' of consciousness, the 'mind–body' problem, the problem of 'other minds', and many more. These problems have sponsored much philosophical reflection and many words but—at present—there appears to be no consensus on the solutions.

We find ourselves in quite a curious situation. Consciousness is very personal. You know what it feels like to have conscious experiences of the external world and you have what I might call an inner mental life. You have thoughts, and you think about these thoughts. You know what your own consciousness *is* or at least what it feels like. So what's the problem?

To answer this question it's helpful to trace the physical processes involved in the conscious perception of a red rose. Now, roses are red because their petals contain a subtle mixture of chemicals called anthocyanins, their redness enhanced if grown in soil of modest acidity. Anthocyanins in the rose petals interact with sunlight, absorbing certain wavelengths and reflecting predominantly red light, electromagnetic radiation with wavelengths between about 620 and 750 billionths of a metre, sitting at the long-wavelength end of the visible spectrum, sandwiched between invisible infrared and orange. Of course, light consists of photons but, no matter how hard we look, we will not find an inherent property of 'redness' in photons with this range of wavelengths. Aside from differences in wavelength (and hence energy, according to the Planck–Einstein relation), there is nothing in the physical properties of photons to distinguish red from green or any other colour.

We keep going. We can trace the chemical and physical changes that result from the interactions of photons with cone cells in your retina all the way to the stimulation of your visual cortex at the back of your brain. Look all you like, but you will not find the *experience* of the colour red in any of this chemistry and physics. It is obviously only when this information is somehow synthesized by your visual cortex do you have a conscious experience of a beautiful red rose.

Just *how* is this supposed to work? This is the hard problem, as philosopher and cognitive scientist David Chalmers explained:[12]

> The really hard problem of consciousness is the problem of *experience*...When we see, for example, we *experience* visual sensations: the felt quality of redness, the experience of dark and light, the quality of depth in a visual field. Other experiences go along with perception in different modalities: the sound of a clarinet, the smell of mothballs. Then there are bodily sensations, from pains to orgasms; mental images that are conjured up internally; the felt quality of emotion, and the experience of a stream of conscious thought. What unites all of these states is that there is something it is like to be in them. All of them are states of experience.

The problem is hard because we not only lack a physical explanation for how this is supposed to happen, we don't even really know how to state the problem properly.

Okay. If the 'how' problem is too hard, can we at least generate some clues by pondering on *where* these experiences might be happening?

The French philosopher René Descartes is rightly regarded as the father of modern philosophy. In his *Discourse on Method*, first published in 1637, he set out to build a whole new philosophical tradition in which there could be no doubt about the absolute truth of its conclusions. From absolute truth, he argued, we obtain certain knowledge. However, to get at absolute truth, he felt he had no choice but to reject as being absolutely false everything in which he could have the slightest reason for doubt. This meant rejecting all the information about the world that he received through his senses.

Why? Well, first, he could not completely rule out the possibility that his senses would deceive him from time to time as, for example, through optical illusions or the sleight of hand and mental manipulations involved in magic tricks.[13] Second, he could not be certain that his perceptions and experiences were not part of some elaborate dream. Finally, he could not be certain that he was not the victim of a wicked demon or evil genius with the ability to manipulate his sensory inputs to create an entirely false impression of the world around him (just like the machines in *The Matrix*).

But he felt that there was at least one thing of which he could be certain. He could be certain that he was a being with a conscious mind that has thoughts. He argued that it would seem contradictory to hold the view that, as a thinking being, he does not exist. Therefore, his own existence was also something about which he could be certain. *Cogito ergo sum*, he concluded. I think therefore I am.

The external physical world is vague and uncertain, and may not appear as it really is. But the conscious mind seems very different. Descartes went on to reason that this must mean that the conscious mind is separate and distinct from the physical world and everything in it, including the unthinking machinery of his body, and his brain. Consciousness must be something 'other', something unphysical.

This mind–body dualism (sometimes called Cartesian dualism) is entirely consistent with belief in the soul or spirit. The body is merely a shell, or host, or mechanical device used for giving outward expression and extension to the unphysical thinking substance. It seems reasonably clear that this kind of dualism is what both von Neumann and Wigner had in mind when they identified consciousness as the place (component **III**) where physical mechanism is no longer applicable, something *outside* the calculation and therefore the ideal place for the collapse of the wavefunction.

But to conclude from this that the conscious mind must therefore be unphysical involves a rather bold leap of logic, one that many contemporary philosophers and neuroscientists believe is indefensible. The trouble is that by disconnecting the mind from the brain and making it unphysical we push it beyond the reach of science and make it completely inaccessible. Science simply can't deal with it. In *The Concept of Mind*, first published in 1949, the philosopher Gilbert Ryle wrote disparagingly of Cartesian-style mind–body dualism, referring to it as the 'ghost in the machine'.[14] In his 1991 book *Consciousness Explained*, the philosopher Daniel Dennett argued that 'accepting dualism is giving up'.[15]

Faced with this impasse, the only way to progress is to make some assumptions. We assume that, however it works, consciousness arises as a direct result of the neural chemical and physical processes that take place in the brain. Our experience of a red rose has a *neural correlate*—it corresponds to the creation of a specific set of chemical and physical states involving a discrete set of neurons located in various parts of the brain. In philosophical terms, this is known as 'materialism'.

Neuroscientists have access to a battery of technologies, such as functional magnetic resonance imaging (fMRI) and positron emission tomography (PET), which can probe the workings of the brain in exquisite detail in non-invasive ways. Experiencing something or thinking about something stimulates one or more parts of the brain. As these parts get to work, they draw glucose and oxygen from the bloodstream. An fMRI scan shows where the oxygen is being concentrated, and so which parts of the brain are 'lighting up' as a result of some sensory stimulus, thought process, emotional response, or memory. A PET scan makes use of a radioactive marker in the bloodstream but otherwise does much the same thing, though with lower resolution.

Neuroscience in its modern form was established only in the second half of the past century, and our understanding has come an awfully long way in that relatively short time. But we must once again acknowledge that whilst studying the brain has revealed more and more of the materialist mechanism, it hasn't yet solved the 'hard problem'.

Some neuroscientists are nevertheless convinced that consciousness is to be found in chemical and neurophysiological events, that consciousness is not a 'thing' but rather an emergent consequence of a complex set of processes occurring in a developed brain.[16]

Grounding consciousness in neuronal activity implies that it is not the exclusive preserve of human beings. In July 2012, a prominent international group of cognitive neuroscientists, neuropharmacologists, neurophysiologists, neuroanatomists, and computational neuroscientists met together at the University of Cambridge in England. After some deliberations, they agreed the Cambridge Declaration on Consciousness, which states:[17]

> the weight of evidence indicates that humans are not unique in possessing the neurological substrates that generate consciousness. Nonhuman animals, including all mammals and birds, and many other creatures, including octopuses, also possess these neurological substrates.

Are cats conscious? The Cambridge Declaration would suggest that they are. Schrödinger's cat might again be spared the discomfort of being both alive and dead, its fate already decided (by its own consciousness) before you lift the lid of the box, and look. To some extent, this answers Bell's challenge. You don't need a PhD to collapse the wavefunction. But you do need to be awake.

According to the current broad consensus, the human mind is the result of evolutionary selection pressures, driven in *Homo sapiens* by feedback loops established between an expanding neural capacity, genetic adaptations, and anatomical changes promoting the development of language capability, and the construction of societies. This is the *social brain* hypothesis. Paleoanthropologists date the specifically human 'light bulb moment' to between 40,000 to 50,000 years ago. This is the moment of the *Great Leap Forward*, or the 'human revolution', a flowering of human innovation and creativity involving the transition to what is known as *behavioural modernity*.

In this 'standard model', consciousness is a natural consequence of the physical, chemical, and biological processes involved in evolution. Now, the bulk properties of water (freezes at 0°C, boils at 100°C) are consequences of the physical properties of up and down quarks, gluons, and electrons, though we would be very hard pressed to predict the former based on what we currently know about the latter. So consciousness is a (so far unpredicted and unpredictable) consequence of the conventional material content of the Universe which emerges when we connect billions of neurons together to form an extended network in the brain, and then run billions of complex computations on this. Consciousness didn't somehow 'pre-exist'.

But what if the Universe has always contained physical events that are, in some sense, 'atoms' of consciousness that existed long before biology? What if one consequence of evolution has then been to assemble these 'atomic' events, orchestrate them, and couple them to activity occurring *within* the neurons in the brain, resulting in what we identify as consciousness? What if the events in question are associated with the distinctly *non-computable* collapse of the wavefunction? Then we have what Penrose and Stuart Hameroff, professor of anesthesiology at the University of Arizona, have called *orchestrated objective reduction*, or Orch-OR, a proposal for a quantum basis for consciousness.

This idea dates back to the early 1990s, and was initially developed separately by Penrose and Hameroff before they chose to combine their efforts in a collaboration. Not surprisingly, Penrose approached the problem from the perspective of mathematics. In his book *The Emperor's New Mind*, he argued in favour of a fundamental role for consciousness in the human comprehension of mathematical truth, one that goes beyond computation. 'We must "see" the truth of a mathematical argument to be convinced of its validity,' he wrote. 'This "seeing" is the very essence of consciousness.'[18] This is entirely consistent with von Neumann's assertion that consciousness 'remains outside of the calculation'.

We've already seen in the previous chapter that, in this same book, Penrose put forward arguments in favour of a role for local mass–energy density and the curvature of spacetime in collapsing the wavefunction, in what was later to become known as Diósi–Penrose theory. However, having convinced himself that consciousness is at heart the result of some kind of non-computable process, and having also proposed a mechanism for the collapse of the wavefunction, he wasn't yet able to make the connection. What he lacked was a physical mechanism that would allow quantum events somehow to govern or determine brain activity, and thence consciousness:[19]

> One might speculate, however, that somewhere deep in the brain, cells are to be found of single quantum sensitivity. If this

proves to be the case, then quantum mechanics will be significantly involved in brain activity.

Hameroff knew where to look. With Richard Watt from the University of Arizona's Department of Electrical Engineering, in 1982 he had hypothesized a role for certain protein polymers called *microtubules* in processing information in the brain. These structures sit *inside* all complex cell systems, including neurons.[*] The polymers self-assemble, allowing the formation of synaptic connections between neurons, and helping to maintain and regulate the strengths of these connections to support cognitive functions. Hameroff and Watt theorized that the polymer subunits (globular proteins called *tubulins*) undergo coherent excitations, forming patterns which support the processing of information much like transistors in a computer.

The conventional wisdom is that information processing in the brain is based on switching between synapses. There are, on average, about 1000 synapses per neuron, each capable of 1000 switching operations per second. The average human brain has about one hundred billion neurons, and hence a capacity of about 10^{17} computational operations per second. But there are 10 *million* tubulin subunits in every cell, capable of switching a million times faster, producing 10^{16} operations per second per neuron. If the information processing really does occur here, then this suggests an enhancement in the number of operations per second to 10^{27}, an increase of *ten orders of magnitude*.

The tubulin subunits possess two distinct lobes, each consisting of about 450 amino acids. Each subunit can adopt at least two different 'conformations'—two different arrangements of its atoms in space—with slightly different distributions of electron

---

[*] They have also been found in some simple prokaryotic cells.

density which generate weak, long-range, so-called 'van der Waals' forces between neighbouring units. These forces are thought to be important in facilitating the switching between conformations, which is the basis of the computational operation.

From the beginning, Hameroff was convinced of the relationship between microtubules and consciousness, not least because of the efficacy of a wide range of very different anaesthetics in temporarily suspending it. Together with Watt, in 1983 he proposed that the chemical anaesthetic seeps into the neuron, disrupting the van der Waals forces between the tubulin subunits, shutting down the computational operations and hence the consciousness of the patient.

On reading *The Emperor's New Mind*, Hameroff approached Penrose and they agreed to collaborate. By the time of publication of Penrose's sequel, *Shadows of the Mind*, in 1994, the Penrose–Hameroff Orch-OR theory was firmly established.

Each tubulin subunit measures about 8 × 4 × 4 billionths of a metre. These link together in polymeric chains which form columns, and 13 columns wrap around to form a hollow tube—the microtubule. These in turn combine with a network of interlinking filaments to make up the neuron's physical support structure, called the cytoskeleton.

The Orch-OR mechanism involves the formation of quantum superpositions of the different tubulin conformations, as depicted in Figure 16a. The subunits interact with their neighbours in a cooperative fashion, enabling the development of extended, coherent superpositions across the microtubule. This is shown as steps 1–6 in Figure 16b, where for clarity the microtubule has been unrolled and flattened out. The individual subunit superpositions are shown as the grey elements in this picture. As the extended superposition builds, it passes a threshold determined by the local mass density (and hence local spacetime curvature) according to the Diósi–Penrose theory. The extended wavefunction collapses

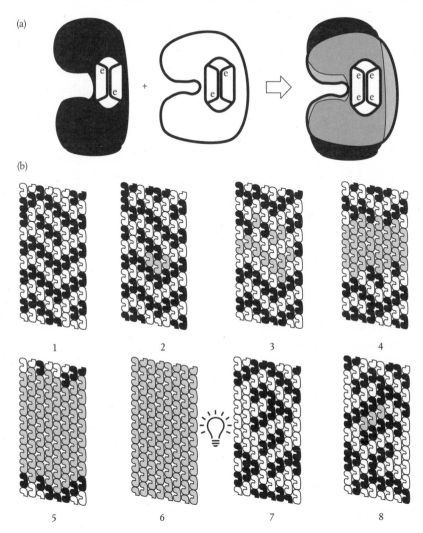

Figure 16 The Penrose–Hameroff Orch-OR theory is based on the idea that tubulin subunits in the polymer chains that make up microtubules inside neurons can enter a superposition of different conformational states, (a). The subunits interact with their neighbours, and a coherent superposition develops across the microtubule, shown in (b) as steps 1–6. When the superposition reaches a critical mass density, the wavefunction collapses (steps 6 to 7), contributing to a conscious experience.

and the tubulin subunits revert to their classical states. This is the transition between steps 6 and 7 and, according to the Orch-OR theory, this is where consciousness happens (hence the light bulb). The process then begins all over again.

This much simplified description of the mechanism doesn't really do it justice. But I think you should get the sense that this is all *very* speculative. It is based on the fusion of ideas from the fringes of quantum physics and neuroscience (and, for that matter, from philosophy). And, unsurprisingly, it has been strongly criticized by both physicists and neuroscientists.

Perhaps the most obvious issue concerns the sustainability of coherent quantum superpositions over what are very large bio-molecular structures. As we saw in Chapter 8, superpositions involving structures intermediate between quantum and classical have been created successfully in the laboratory, including organic molecules containing up to 430 atoms. Whilst it's fair to say that we do not yet know what the upper limit might be, the larger the system, the more difficult it is to protect it from the effects of environmental decoherence. This is why MAQRO is a space mission.

But each tubulin subunit is a protein structure containing over *ten thousand atoms*.[20] Microtubules vary in length from about 200 up to 25,000 billionths of a metre. The shorter length implies a polymer column of just 25 subunits, and 13 columns implies a microtubule consisting of 325 subunits in total. The Orch-OR mechanism then calls for a coherent quantum superposition spanning a structure containing on the order of 325,000 atoms, and which must be sustained for millisecond timescales before collapsing. Contrast this with the macroscopic objects suggested for the MAQRO mission, which are small 'nanospheres' with diameters of about 100 billionths of a metre.

It seems extremely unlikely that a coherent quantum super-position can be sustained in the kind of 'warm, wet, and noisy' environment likely to be typical of neurons in a working brain. In 2000, theorist Max Tegmark argued that decoherence time-scales on the order of a tenth of a trillionth (10^{-13}) to a hundredth of a millionth of a trillionth (10^{-20}) of a second are more likely in this kind of environment.[21]

But, once again, we must never underestimate the power of a realistic interpretation to inspire and motivate interest, consistent with Proposition #4. Despite its very speculative nature, the Penrose–Hameroff Orch-OR theory has many components that are potentially accessible to experiment and arguably makes many testable predictions. In a recent 2014 updating of the theory and review of its status, Hameroff and Penrose examine how 20 predictions they had offered in 1998 have fared in the interim. They drew much comfort from a recent discovery by the research group led by Anirban Bandyopadhyay at the National Institute of Material Sciences in Japan, of memory-switching in a single brain microtubule.[22] They concluded that the theory had actually fared rather well, giving a 'viable scientific proposal aimed at providing an understanding of the phenomenon of consciousness'.[23]

Needless to say, the one component of the theory for which there is as yet no empirical evidence is the Diósi–Penrose OR mechanism. For now, any potential role for the non-computable collapse of the wavefunction in facilitating consciousness remains stranded on the beaches of Metaphysical Reality.

Even if evidence for a connection between quantum mechanics and consciousness can one day be found, Chalmers argues that this will still not solve the hard problem: 'when it comes to the explanation of experience, quantum processes are in the same boat as any other. The question of *why* these processes should give rise to experience is entirely unanswered.'[24]

10

Quantum Mechanics is Incomplete
BECAUSE...
OKAY, I GIVE UP

The View from Charybdis: Everett, Many Worlds, and the Multiverse

Y ou will gather from the title I've chosen for this final chapter that I'm not particularly enamoured of the interpretations we will consider here. Whilst this is certainly true, my ambition is nevertheless to make the case both in favour and against these interpretations, as best I can, so that you can judge for yourself. I'll explain my problems with them as we go.

Irrespective of what I think, it has never ceased to amaze me that one of the simplest solutions to the problem of the collapse of the wavefunction leads to one of the most bizarre conclusions in all of quantum mechanics, if not all of physics. This solution involves first recognizing that we have absolutely no evidence for the collapse. It was introduced by von Neumann as a postulate. We have seen quantum systems in various stages of coherence, and we can create superpositions of objects that are intermediate between quantum and classical, but we have to date never *seen* a system undergo collapse, nor have we seen superpositions of large, cat-sized objects. In any realistic interpretation of the wavefunction, the notion of collapse has always been nothing more than a rather ad hoc device to get us from a system we are obliged to describe as a superposition of possible measurement outcomes to a system with a single outcome.

There's another reason to question the need for the collapse, as I already mentioned in passing in Chapter 6. This is related to the relationship between quantum theory and the spacetime described by Einstein's general theory of relativity.

Einstein presented the general theory in a series of lectures delivered to the Prussian Academy of Sciences in Berlin, culminating in a final, triumphant lecture on 25 November 1915. Yet within a few short months he was back, advising the Academy that his new theory of gravitation might need to be modified: 'the quantum theory should, it seems, modify not only Maxwell's electrodynamics but also the new theory of gravitation'.[1]

Attempts to construct a quantum theory of gravity were begun in 1930 by Bohr's protégé, Leon Rosenfeld. As I explained in Chapter 5, there are three 'roads' that can be taken, one of which involves the quantization of spacetime in general relativity and leads to structures such as loop quantum gravity. The result is a quantum theory of the gravitational field, a quantum theory of spacetime itself.

The difficulties are not to be underestimated. Quantum mechanics is formulated against an assumed background spacetime, conceived not much differently from the absolute space and time of Newton's classical mechanics. Quantum mechanics is 'background-dependent'. In contrast, the spacetime of general relativity is dynamic. The geometry of spacetime is variable: it *emerges* as a consequence of physical interactions involving mass–energy. General relativity is 'background-independent'.

In any realistic interpretation of the wavefunction, these different ways of conceiving of space and time give us a real headache.

In quantum mechanics, we routinely treat a quantum system as though it is sitting in a box, isolated from an outside world that, for the sake of simplicity, we prefer not to consider. Inside this box, we apply the Schrödinger equation and the wavefunction of the system evolves smoothly and continuously as it moves in time in a distributed, non-local fashion from place to place. This is Axiom #5, or von Neumann's process 2. Satisfied, we now turn our attention to the classical measuring device, which is located in the world outside the box. When this interacts with the quantum system we suppose that the wavefunction collapses according to process 1.

Just how are we meant to apply this logic to a quantum spacetime? Aside from some 'spooky' non-local effects, we can regard a quantum system formed from a material particle or ensemble of particles as being broadly 'here', in this place in the Universe, and therefore inside this box. The box is clearly defined by boundaries imagined *in* spacetime. But if we consider the entirety of spacetime itself, no such boundaries can be imagined. A quantum theory of spacetime is, kind of by definition, a quantum theory of the entire Universe, or a quantum cosmology.

In any realistic interpretation of the wavefunction, the need to invoke a separate process for 'measurement' really makes quite a mess of things, as this necessarily assumes a perspective that is outside of the system being measured and, so far as we know, there can be nothing outside the Universe. This dilemma caught the attention of Hugh Everett III, a chemical engineering graduate who had migrated first to mathematics (including military game theory) at Princeton University, and then in 1955 to studies for a PhD in physics under the supervision of John Wheeler. In a 1957 paper based on his dissertation, he wrote:[2]

No way is it evident to apply the conventional formulation of quantum mechanics to a system that is not subject to external observation. The whole interpretative scheme of that formalism rests upon the notion of external observation.

Given the problems that the collapse creates and the lack of any direct evidence for it, why not simply get rid of it? I've already explained that scientists down the centuries have made a useful habit of eliminating from their theories all the unnecessary baggage and frills, typically introduced to satisfy certain metaphysical preconceptions but ultimately unsupported by the empirical facts. This is what Everett chose to do. In his dissertation, he followed von Neumann's logic in assuming that the quantum mechanics of process 2 applies equally to large-scale, classical objects, and offered the following alternative interpretation:[3]

> To assume the universal validity of the quantum description, *by the complete abandonment of Process 1.* The general validity of pure wave mechanics, without any statistical assertions, is assumed for all physical systems, including observers and measuring apparata. Observation processes are to be described completely by the [wave]function of the composite system which includes the observer and his object-system, and which at all times obeys the [Schrödinger] wave equation (Process 2).

At first sight, this suggestion seems somewhat counterproductive. If it is indeed our experience that pointers point in only one direction at a time and cats are either alive or dead, then giving up the collapse would seem to be taking us in the wrong direction. Surely, we are confronted by Schrödinger's infinite regress, with an endless complexity of superpositions of measuring devices, cats, and ultimately human observers?

But, of course, we never *experience* superpositions of large-scale, classical objects. Everett argued that the only way out of this contradiction is to suppose that *all* possible measurement outcomes are realized.

How can this be? Once again, it's easier to follow Everett's logic using an example so, for the last time, let's return to our quantum particle A prepared in a superposition of ↑ and ↓ spin states. The total wavefunction encounters a measuring device, and the larger system evolves smoothly into a superposition of the outcomes $A_\uparrow$ and $A_\downarrow$. These trigger a response from a gauge attached to the measuring device, entangling the outcomes with the 'pointer states' of the gauge, as before, resulting in a superposition of the product states $A_\uparrow$🔒 and $A_\downarrow$🔒. But this is no longer the kind of superposition we've been considering thus far. In his dissertation, Everett wrote:[4]

> Whereas before the observation we had a single observer state afterwards there were a number of different states for the observer, all occurring in a superposition. Each of these separate states is a state for an observer, so that we can speak of the different observers described by the different states. On the other hand, the same physical system is involved, and from this viewpoint it is the *same* observer, which is in different states for different elements of the superposition (i.e. has had different experiences in the separate elements of the superposition).

Everett was not proposing that the observer enters some kind of conscious superposition, in which both outcomes are experienced simultaneously, but rather that the observer 'splits' between different states. In our example, one of these observer states corresponds to the experience of $A_\uparrow$ and another corresponds to the experience of $A_\downarrow$:

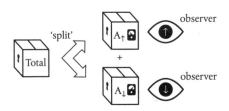

Everett was not clear on the nature or cause of the 'split', but there is evidence in his writings that he interpreted it quite literally as a physical phenomenon acting on (or promoted by) a real wavefunction. Wheeler was much more cautious, annotating one of Everett's manuscripts with the comment 'Split? Better words needed', and recommending that Everett rephrase his arguments to avoid 'mystical misinterpretations by too many unskilled readers'.[5]

This is Everett's 'relative state' interpretation, the use of the word 'relative' deriving from the correlations established between the components of the superposition in the total wavefunction and the different experiences of the observer following the split. One observer state corresponds to $A_\uparrow$ and the other to $A_\downarrow$ *relative* to the first. Everett argued that quantum probability is then nothing more than a *subjective* probability imposed by an observer who, in a succession of observations on identically prepared A particles, notes that the outcomes are random, with a 50:50 probability of observing $\uparrow$ or $\downarrow$. Everett wrote: 'We are then led to the novel situation in which the formal theory is objectively continuous and causal, while subjectively discontinuous and probabilistic.'[6]

Despite his concern about some of Everett's phraseology, Wheeler sent his dissertation to Bohr and later visited Bohr in Copenhagen in an effort to drum up support for what he judged to be a promising approach. These efforts culminated in a visit by Everett himself in March 1959. But all was in vain. Bohr, ever mindful of the use and misuse of language in physical descriptions, had no wish even to discuss 'any new (strange) upstart theory'.[7] Everett's approach challenged too many elements of the Copenhagen orthodoxy, such as the complementarity of waves and particles and the interpretation of quantum probability. The Copenhagen school rejected outright any suggestion that quantum mechanics could be applied to classical objects.

Everett was dismayed. By this time he had already been working for two and a half years with the Pentagon's Weapons Systems
Evaluation Group, and he rebuffed Wheeler's attempts to lure
him back to academia. He was invited to present his relative state
interpretation at a conference in 1962, but his ideas were largely
ignored by the physics community at the time.

With one notable exception.

Bryce DeWitt had been intrigued by Everett's analysis and
had written to him in 1957 with a lengthy critique. If the observer state splits every time an observation is made then why,
DeWitt wondered, is the observer not aware of this? Everett
responded with an analogy. We have no reason to question the
conclusions of astronomy and classical physics which tell us
that the Earth spins on its axis as it orbits the Sun. And yet
because of our own inertia, we don't directly experience this
motion. Likewise, an observer maintains a sense of a single
identity and a single history that can be reconstructed from
memories, unaware that there exist multiple versions of him or
herself, all with different recollections of the sequence of events.
Everett slipped this argument into the proofs of his 1957 paper as
a footnote.

DeWitt was convinced. Struggling to come to terms with a
quantum cosmology seemingly at odds with a formalism that
appeared to place far too much emphasis on external 'measurement', DeWitt sought to raise the profile of Everett's interpretation in a paper published by *Physics Today* in September 1970.
Bohr had died in 1962, and there might have been signs that the
Copenhagen interpretation was starting to lose its stranglehold.
DeWitt may have judged that the time for pussyfooting around
with politically correct language was over, and he chose to describe
the interpretation using words that the old guard (including
Wheeler) would never have approved.

He wrote: 'I shall focus on one [interpretation] that pictures the universe as continuously splitting into a multiplicity of mutually unobservable but equally real worlds, in each one of which a measurement does give a definite result.'[8] Thus Everett's relative state formulation became the *many-worlds interpretation* of quantum mechanics. Almost overnight, Everett's interpretation transformed from being one of the most obscure to one of the most controversial.

Everett appears to have been pleased with DeWitt's choice of words.[*] Wheeler once advised him that, whilst he 'mostly' believed in the interpretation, he reserved Tuesdays once a month to disbelieve it. But his reservations were stronger than this anecdote would imply, and in time he withdrew his support, such as it had been, later accepting that the interpretation 'carried too much metaphysical baggage along with it', and made 'science into a kind of mysticism'.[9] But even as he distanced himself he acknowledged Everett's contribution as one of the most original and important in decades.[10]

DeWitt secured a copy of Everett's original dissertation from his wife Nancy. With Everett's blessing and the help of his student Neill Graham, DeWitt published this in 1973, together with Everett's 1957 paper, an 'assessment' of this paper published back

---

[*] Interestingly, in his *Physics Today* paper, De Witt argued that von Neumann's collapse postulate was part of the 'conventional' or 'Copenhagen' interpretation. And yet a formal theory of measurement was never part of the Copenhagen orthodoxy (Bohr's biographer Abraham Pais found an entry in one of his old notebooks pertaining to a lecture delivered by Bohr in November 1954 which reads: '[Bohr] thinks that the notion "quantum theory of measurement" is wrongly put' (see Abraham Pais, *Niels Bohr's Times*, Oxford University Press, 1991, p. 435). Indeed, why would an anti-realist interpretation which draws a line between quantum and classical worlds require a quantum theory of measurement? Whilst 'conventional' interpretation is fair, De Witt misrepresents 'Copenhagen' and—to me at least—it appears that he is seeking to demonize it. More on this later in this chapter.

to back in the same journal by Wheeler, DeWitt's *Physics Today* article, and a couple of further supportive articles by DeWitt, Graham, and Leon Cooper and Deborah van Vechten. The book is titled *The Many Worlds Interpretation of Quantum Mechanics.*

Schrödinger's cat is no longer simultaneously alive and dead in one and the same world, it is alive in one world and dead in another. With repeated measurements, the number of worlds splitting off from each other multiplies rapidly. The act of measurement has no special place in the many-worlds interpretation, so there is no reason to define measurement to be distinct from *any* process involving a quantum superposition. We can suppose that a great many such processes have occurred since the Big Bang origin of the Universe, some 13.8 billion years ago. Each will have split the world into as many branches as there have been components in the various superpositions that have formed since.

In his *Physics Today* article, DeWitt estimated that by now there must be more than 10^{100+} different branches, or distinct worlds. Each of these worlds contain 'slightly imperfect copies of oneself all splitting into further copies, which ultimately become unrecognizable ... Here is schizophrenia with a vengeance.'[11]

Because it assumes *only* the continuous mechanics described by the Schrödinger equation, and nothing more, the many-worlds interpretation is often advertised as satisfying the demand that quantum mechanics be regarded as a complete theory. If that's the case, then the title I've chosen for this chapter is incorrect. Needless to say, I don't agree. Whilst the theory might be mathematically complete (and there are more arguments to come on this score), I think completing the theory by invoking an enormous multiplicity of mutually unobservable real worlds is rather costly, and certainly not 'nothing'. It has been observed that the many-worlds interpretation is 'cheap on assumptions, but expensive on universes'.[12]

Caught in a strong trade wind, the Ship of Science is hurtling relentlessly towards Charybdis, that dangerous whirlpool of metaphysical nonsense about the nature of reality.

There have been many efforts at rehabilitation in the years that have followed. It's not clear what splitting the world implies for the interference terms in the superposition, and one of Dieter Zeh's motivations for developing ideas about decoherence was to replace the notion of 'many worlds' with that of 'multiconsciousness', or what is now known as the 'many minds' interpretation.[13] The splitting or branching of worlds or universes is replaced by the splitting or branching of the observer's consciousness. The observer is never consciously aware of the superposition, since environmental decoherence destroys (or, more correctly, suppresses) the interference terms. The observer is only ever consciously aware of one outcome, but the other outcomes nevertheless exist in the observer's mind, although these alternative states of consciousness are inaccessible. In short, the observer does not possess a single mind but rather a multiplicity (or even a continuous infinity) of minds, each weighted according to the amplitude of the wavefunction such that one is dominant.

Space precludes a more detailed discussion of the many minds interpretation, a term coined by philosophers David Albert and Barry Loewer in 1988. The relationship between the mind–body problem and quantum mechanics was independently explored by philosopher Michael Lockwood, in his 1989 book *Mind, Brain & the Quantum: The Compound 'I'*.[14] The many minds interpretation is presented in Albert's popular book *Quantum Mechanics and Experience*, first published in 1992.[15]

John Bell saw close similarities between many worlds and de Broglie–Bohm theory, arguing that Everett's original interpretation could be rationalized as the pilot wave theory but without particle paths. In this sense, the splitting or 'branching' implied by many worlds is no more or less extravagant than the 'empty waves' of de Broglie–Bohm theory. Like Wheeler, Bell regarded the endless multiplication of worlds to be rather over the top, arguing that it serves no real purpose and so can be safely eliminated. The novel element that Everett had introduced was 'a repudiation of the concept of the "past"'.[16] Instead of a branching system of worlds sprouting like the limbs of a tree, Bell suggested instead that the various particle 'histories' run in parallel, sometimes coalescing to produce interference effects. The outcomes are then determined by summing over these histories, with no association discernible between any particular present and any particular past.

But this 'neo-Everettian' line of reasoning really just locks us into an endless debate about trading off different metaphysical preconceptions. If you personally prefer to stick with the simplicity of a real universal wavefunction whose motion is fully described by process 2, then you must decide whether you're ready to accept a multiplicity of worlds, defined either as physically real or simply 'effective', for all practical purposes. You can avoid this by resorting to de Broglie–Bohm theory, but then as we've seen you then need to reconcile yourself to non-local spooky action at a distance.

Zeh introduced decoherence into the mix as a way of sharpening the relationship between quantum and classical systems. Bell talked about replacing many worlds with many particle 'histories'. It's therefore a relatively short step from this back to the decoherent histories interpretation which I presented in Chapter 6. In his popular book *The Quark and the Jaguar*, published in 1994, Murray Gell-Mann explained it like this:[17]

We believe Everett's work to be useful and important, but we believe that there is much more to be done. In some cases too, his choice of vocabulary and that of subsequent commentators on his work have created confusion. For example, his interpret-ation is often described in terms of 'many worlds', whereas we believe that 'many alternative histories of the universe' is what is really meant. Furthermore, the many worlds are described as being 'all equally real', whereas we believe it is less confusing to speak of 'many histories, all treated alike by the theory except for their different probabilities'.

This is fine, as far as it goes. But note that this change of perspec-tive isn't only about reinterpreting the words we use. The many-worlds interpretation is fundamentally realist—it assumes the existence of a real universal wavefunction and possibly a multi-plicity of real worlds—whereas, as we saw in Chapter 6, decoher-ent histories is broadly anti-realist: 'all equally real' is diluted to 'all treated alike by the theory'. You begin to get the sense that there's no easy way out.

And herein lies the rub. Many theoretical physicists and philo-sophers who advocate the many-worlds interpretation, or who claim to be 'Everettians' or 'neo-Everettians', don't necessarily buy into a single interpretation or a single set of metaphysical pre-conceptions.* 'After 50 years, there is no well-defined, generally agreed set of assumptions and postulates that together constitute "the Everett interpretation of quantum theory",' wrote Adrian Kent in 2010.[18] The chances are that advocates of many worlds buy into their own, possibly individually unique, understanding of what this actually means for them. This is important for what follows, as my criticism is confined to those theorists and philo-sophers who have not only embraced their inner metaphysician, but who have decided to go all-in with the metaphysics.

---

* Just as there's no single 'Copenhagen interpretation'.

In May 1977, DeWitt and Wheeler invited Everett to participate in a conference organized at the University of Texas in Austin. The subjects of the conference were human consciousness and computer-generated artificial intelligence, likely reflecting Wheeler's growing interest in the role of consciousness in helping define the laws of physics in a 'participatory universe'. Everett gave a seminar not on many worlds, but on machine self-learning.

During a lunchtime break at a beer-garden restaurant, DeWitt arranged for Everett to sit alongside one of Wheeler's young graduate students. Over lunch the student probed Everett about his interpretation, and about how he himself preferred to think about it. Although Everett's career had by now taken him far from academia and he was no longer immersed in questions about the interpretation of quantum mechanics, the student was very impressed. Everett was still very much in tune with the debate.

The student's name was David Deutsch.

Deutsch would go on to develop his own singular version of the many-worlds interpretation. He argued that the notion of a universe 'branching' with each and every transition involving a quantum superposition couldn't be right. The simple fact that interference is possible with a *single particle* tells us that reality consists of an infinity of *parallel universes*, which form what is now generally known as the *multiverse*.

To follow Deutsch's arguments let's return to the description of two-slit interference involving electrons in Chapter 1, and especially Figure 4. Look again at Figure 4a, in which we see just a few scattered points of brightness each indicating that 'an electron struck here'. In Chapter 1, I explained that interference effects with single electrons arguably demonstrate that each individual electron behaves as a wave—conceived of as a real wave or a 'wave of information' (whatever that means)—passing through both slits at once. But what if electrons really do maintain their integrity as real, localized particles, capable of passing through only one slit or the other? Deutsch argues that the only way to recover interference from this is to propose that each electron is accompanied by a host of 'shadow' or 'ghost' electrons, which pass through both slits and interfere with the path of the visible electron.

Whilst these 'shadow' electrons clearly influence the path of the visible electron, they are themselves not detectable—they make no other tangible impression. One explanation for this is that the 'shadow' electrons do not exist in 'our' universe. Instead they inhabit 'a huge number of parallel universes, each similar in composition to the tangible one, and each obeying the same laws of physics, but differing in that the particles are in different positions in different universes'.[19] When we observe single-particle interference, what we see is not a quantum wave–particle interfering with itself, but rather a host of particles in parallel universes interfering with a particle in our own, tangible universe.

In this interpretation, the 'tangible' universe is simply the one which you experience and with which you are familiar. It is not privileged or unique: there is no 'master universe'. In fact, there is a multiplicity of 'you' in a multiplicity of universes, and each regards their universe as the tangible one. Because of the quantum nature of the reality on which these different universes are founded, some of these 'you's have had different experiences and have different histories or recollections of events. As Deutsch explained: 'Many of those Davids are at this moment writing these very words. Some are putting it better. Others have gone for a cup of tea.'[20]

This is quite a lot to swallow, particularly when we consider that we have absolutely no empirical evidence of the existence of all these parallel universes. But Deutsch argues that the multiverse interpretation is the only way we can explain the extraordinary potential for *quantum computing*.

This is worth a short diversion.

The processors that sit in every desktop computer, laptop, tablet, smartphone, smartwatch, and item of wearable technology perform their computations on strings of binary information called 'bits', consisting of '0's and '1's. Now, classical bits have the values 0 *or* 1. Their values are one or the other. They can't be added together in mysterious ways to make superpositions of 0 *and* 1. However, if we make bits out of quantum particles such as photons or electrons, then curious superpositions of 0 and 1 become perfectly possible. For example, we could assign the value '0' to the spin state ↑ and the value '1' to the spin state ↓. Such 'quantum bits' are referred to as 'qubits'. Because we can form superpositions of qubits, the processing of quantum information works very differently compared with the processing of classical information.

A system of classical bits can form only *one* 'bit string' at a time, such as 01001101. But in a system consisting of qubits we can form superpositions of all the different possible combinations. The physical state of the superposition is determined by the amplitudes of the wavefunctions of each qubit combination, subject to the restriction that the squares of these amplitudes sum to 1 (measurement can give one, and only one, bit string).

And here's where it gets very interesting. If we apply a computational process to a classical bit, then the value of that bit may change from one possibility to another. For example, a string with eight bits may change from 01001101 to 01001001. But applying a computational process to a qubit superposition changes *all* the components of the superposition *simultaneously*. An input superposition yields an output superposition.

A computation performed on an input in a classical computer is achieved by a few transistors which turn on or off. A linear input gives us a linear output, and if we want to perform more computations within a certain time we need to pack more transistors into the processor. We've been fortunate to witness an extraordinary growth in computer power in the past 30 years.[*] The Intel 4004, introduced in 1971, held 2,300 transistors in a processor measuring 12 square millimetres. The Apple 12 Bionic, used in the iPhone XS, XS Max, and XR, which were launched in September 2018, packs about 7 *billion* transistors into a processor measuring 83 square millimetres. The record is currently on the order of 20 to 30 billion transistors, depending on the type of chip.

But this is nothing compared with a quantum computer that promises an *exponential* amount of computation in the same amount of time.

The cryptographic systems used for most Internet-based communications and financial transactions are based on the simple fact that finding the prime factors of very large integer numbers requires a vast amount of computing power.[†] For example, it has been estimated that a network of a million conventional computers would require over a million years to find the prime factors of a 250-digit number. Yet this feat could, in principle, be performed in a matter of minutes using a single quantum computer.[‡]

---

[*] According to *Moore's law* (named for Gordon Moore, one of the founders of Intel), the number of transistors in a computer processor doubles every 18 months to 2 years.

[†] Factorization involves finding integer numbers that are factors of a larger number. For example, the number 42 can be factored as 6 × 7. Prime factorization involves finding factors that are prime numbers, numbers that themselves can't be factored any further. Thus, the prime factors of the number 42 are 2, 3, and 7.

[‡] We need to be a bit careful here. Quantum computers offer an exponentially faster processing speed only for certain problems, such as factoring. For other problems—such as sorting—quantum computers offer no advantages at all. See Peter Shor's interview with John Horgan: https://blogs.scientificamerican.com/cross-check/quantum-computing-for-english-majors/

Deutsch claims that this kind of enhancement in computing power is only possible by leveraging the existence of the multiverse. When using a quantum computer to factor a number requires 10^{500} or so times the computational resources that we see to be physically present, where then is the number factorized?[21] Given that there are only about 10^{80} atoms in the visible Universe, Deutsch argues that to complete a quantum computation we need to call on the resources available in a multitude of other parallel universes.

Needless to say, a quantum computer with this kind of capability is not yet available. As I've already mentioned, systems prepared in a quantum superposition are extremely sensitive to environmental decoherence and, in a quantum computer, the superposition must be manipulated without destroying it. It is allowed to decohere only when the computation is complete. To date, quantum computing has been successfully demonstrated using only a relatively small number of qubits.* We may still be 20–30 years away from the kind of quantum computer that could threaten the encryption systems used today.

Practical considerations notwithstanding, we need to address Deutsch's principal assertion—that the existence of the multiverse is the *only* explanation for the enhancement of processing speed in a quantum computer.

It should come as no real surprise to learn that there are more than a few problems with the many-worlds interpretation that carry through to Deutsch's multiverse. First, there is a problem with the way that many worlds handles quantum probability,

---

* The D-Wave 2000Q System is based on a quantum processor consisting of 2,000 superconducting qubits and purportedly costs $15 million. But there's a caveat. Whilst it is generally understood that D-Wave machines are indeed based on quantum computing, many of the qubits are needed for error correction rather than calculation.

and for many years arguments have bounced back and forth over whether it is possible to derive the Born rule using this interpretation. Everett claimed to have solved this problem already in his dissertation, but not everybody is satisfied. There seems nothing to prevent an observer in one particular universe observing a sequence of measurement outcomes that do not conform to Born-rule expectations.

A colourful example was provided by many-worlds enthusiast Max Tegmark (another former student of Wheeler's). He proposed an experiment to test the many-worlds interpretation that is not for the faint of heart. Indeed, experimentalists of weak disposition should look away now.

Instead of connecting our measuring device to a gauge, imagine that we connect it to a machine gun. This is set up so that when particle A is detected to be in a ↓ state, a bullet is loaded into the chamber of the machine gun and the gun fires. If particle A is detected to be in an ↑ state, no bullet is loaded and the gun instead just gives an audible 'click'. We stand well back, and turn on our preparation device. This produces a steady stream of A particles in a superposition of ↑ and ↓ states. We satisfy ourselves that the apparatus fires bullets and gives audible clicks with equal frequency in an apparently random sequence.

Now for the grisly bit.

You stand with your head in front of the machine gun. (I'm afraid I'm not so convinced by arguments for the many-worlds interpretation that I'm prepared to risk my life in this way, and *somebody* has to do this experiment.) Of course, as an advocate of many worlds, you presume that all you will hear is long series of audible clicks. You are aware that there are worlds in which your brains have been liberally distributed on the laboratory walls, but you are not particularly worried by this because there are other worlds where you are spared.

By definition, if you are not dead, then your history is one in which you have heard only a long series of audible clicks. You can check that the apparatus is still working properly by moving your head to one side, at which point you will start to hear gunfire again. If, on the other hand, the many-worlds interpretation is wrong and the wavefunction simply represents coded information, or the collapse is a physically real phenomenon, then you might be lucky with the first few measurements but, make no mistake, you will soon be killed. Your continued existence (indeed, you appear to be miraculously invulnerable to an apparatus that really should kill you) would appear to be convincing evidence that the many-worlds interpretation is right.

Apart from the obvious risk to your life, the problem with this experiment becomes apparent as soon as you try to publish a paper describing your findings to a sceptical physics community. There may be worlds in which all you recorded was a long series of audible clicks. There are, however, many other worlds where I was left with a very unpleasant mess and a lot of explaining to do. The possibility of entering one of these worlds when you repeat the experiment does not disappear, and you will find that you have a hard time convincing your peers that you are anything other than quite mad. Tegmark wrote:[22]

> Perhaps the greatest irony of quantum mechanics is that...if once you feel ready to die, you repeatedly attempt quantum suicide: you will experimentally convince yourself that the [many-worlds interpretation] is correct, but you can never convince anyone else!

I'd like to make a further point. A history in which all you've heard is a long series of audible clicks suggests a world in which the superposition only ever produces an $A_\uparrow$ result. This is much like tossing a fair coin but only ever getting 'heads', or rolling a dice and only getting six. Yet the Born rule insists that there should be a 50:50 probability of observing $A_\uparrow$ *and* $A_\downarrow$. How can the Born rule be recovered from this?

Philosopher David Wallace devotes three long chapters to probability and statistical inference in his book *The Emergent Multiverse: Quantum Theory According to the Everett Interpretation*, published in 2012. This is perhaps the most comprehensive summary of the Everett interpretation available, though readers should note that this is not a popular account, nor is it likely to be accessible to graduate students without some training in both physics and philosophy. Wallace seeks to avoid the connotations of 'many worlds' (the reader can almost hear Wheeler, whispering over his shoulder: 'better words needed!') but, like Everett, Wallace resorts to subjective, Bayesian decision theory, arguing that the components of the wavefunction are translated to different 'weights' for different branches. The observer then subjectively assigns 'probabilities to the outcomes of future events in accordance with the Born Rule'.[23]

But in Tegmark's quantum suicide experiment, you only ever experience a long series of audible clicks, and this would surely lead you to conclude that in future events you're only ever going to get the result $A_\uparrow$. Getting around this challenge requires some interesting mental gymnastics. If this is about the expectation of different subjective experiences, then we might be inclined to accept that death is not an experience. The experiment is flawed precisely because of the way it is set up. Wallace writes: 'experiments which provide the evidential basis for quantum mechanics do not generally involve the death of the experimenter, far less of third parties such as the writer!'[24]

I confess I'm not entirely convinced. And I'm not alone. Lev Vaidman—another many-worlds enthusiast—isn't completely convinced, either.[25]

Look back briefly at the discussion in Chapter 5. Suppose we prepare an ensemble of A particles in a superposition of spin states ↑ and ↓, but we *measure* the outcomes in the basis + and −. With any one of a potentially large number of ways of expressing the superposition in terms of different basis states we can freely choose from, how are the branches supposed to 'know' which basis corresponds to the outcomes that are to be observed? This is the problem of the 'preferred basis'.

As we've seen, the localization of the different measurement outcomes through decoherence arguably helps to get rid of (or, at least, minimize) the interference terms, but the wavefunction must still be *presumed* to be somehow steered towards the preferred basis in the process. In Wallace's version of the Everett interpretation, the real wavefunction interacts with the classical measuring device and the environment, evolving naturally into a superposition of the measurement outcomes. The interference terms are dampened, and the eventual outcomes are realized in different branches. Wallace writes: 'Decoherence is a dynamical process by which two components of a complex entity (the quantum state) come to evolve independently of one another, and it occurs owing to rather high-level, emergent consequences of the particular dynamics and the initial state of our universe.'[26] So, decoherence naturally and smoothly connects the initial wavefunction with the preferred basis, determined by how the experiment is set up.

This might sound quite plausible, but now let's—at last—return to quantum computing. It turns out that there is more than one way to process qubits in a quantum computer. Instead of processing them in sequence, a 'one-way' quantum computer based on a highly entangled 'cluster state' proceeds by feeding forward the outcomes of irreversible measurements performed on single qubits. The (random) outcome from one step determines the *basis* to be applied for the next step, and the nature of the measurements and the order in which they are executed can be configured so that they compute a specific algorithm.* Such a computer was proposed in 2001 by Robert Raussendorf and Hans Briegel,[27] and a practical demonstration of computations using a four-qubit cluster state was reported in 2005.[28]

So, in a cluster state computer, the basis for each measurement changes randomly from one step in the calculation to the next, and will differ from one qubit to the next. Note that whilst 'measurement' here is irreversible, it does not involve an act of amplification, in which we might invoke decoherence. Quite the contrary. The superpositions required for quantum computing must remain coherent—the maximum length of a computation is determined by the length of time that coherence can be maintained. In such a system, decoherence is unwanted 'noise'.

Even if it were possible for a preferred basis somehow to emerge during one step of the computation, this is not necessarily the basis needed for the next step in the sequence. Decoherence can't help us here. Philosopher Michael Cuffaro writes: 'Thus there is no way in which to characterise the cluster state computer as performing its computations in many worlds, for there is no way, in the context of the cluster state computer, to even *define* these worlds for the purposes of describing the computation as a whole.'[29]

In this case, it's doubtful that many worlds or the multiverse serve any useful purpose as a way of thinking about quantum computation. Cuffaro believes that those advocates who take the physical reality of the different worlds or branches rather less seriously should be broadly in agreement with his arguments.[30] In other words, if the many different worlds simply represent a

---

* For example, a specific configuration of four qubits organized in a circuit can be used to compute a database search algorithm devised by Lov Grover in 1996.

useful way of thinking about the problem, but are not assumed to be physically real, then this is no big deal.

Wallace acknowledges that whilst what he calls 'massive parallelism' (i.e. the multiverse) might have been helpful historically as a way of thinking about quantum computation, it 'has not been especially productive subsequently'. He continues: 'Nor would the Everett interpretation particularly lead one to think otherwise: the massively parallel-classical-goings-on way to understand a quantum state is something that occurs only emergently, in the right circumstances, and there's no reason it has to be available for inherently microscopic (i.e. not decoherent) systems.'[31] But, in circumstances where decoherence is possible, as far as Wallace is concerned the many emergent worlds are real 'in the same sense that Earth and Mars are real'.[32]

Wallace is a philosopher, and his musings on the reality of the multiverse are largely confined to philosophy journals and books. But Deutsch is a scientist. Yet he too insists that we accept that these worlds really do exist: 'It's my opinion that the state of the arguments, and evidence, about other universes closely parallels that about dinosaurs. Namely: they're real—get over it.'[33]

I believe we've now crossed a threshold. I've claimed that it is impossible to do science of any kind without metaphysics. But when the metaphysics is completely overwhelming and the hope of any contact with Empirical Reality is abandoned—when the metaphysics is all there is—I would argue that such speculations are *no longer scientific*. Now perched on the very edge of Charybdis, the Ship of Science is caught in its powerful grip. We watch, dismayed, as it starts to slip into the maelstrom.

In recent years, other varieties of multiverse theory have entered the public consciousness, derived from the cosmological theory of 'eternal inflation' and the so-called 'cosmic landscape' of superstring theory. These variants are very different: they were conceived for different reasons and purport to 'explain' different aspects of foundational physics and cosmology. But these variants provoke much the same line of argument. Thus, Martin Rees, Britain's Astronomer Royal, declares that the cosmological multiverse is not metaphysics but exciting science, which 'may be true', and on which he'd bet his dog's life.[34]

Despite their different origin and explanatory purpose, some theorists have sought to conflate these different multiverse theories into a single structure. In his recent book *Our Mathematical Universe*, Tegmark organizes these different approaches into a nested hierarchy of four 'levels'.[35] The Level I multiverse comprises universes with different sets of initial Big Bang conditions and histories but the same fundamental laws of physics. This is the multiverse of eternal inflation. Level II is a multiverse in which universes have the same fundamental laws of physics but different effective laws (different physical constants, for example). We happen to live in a universe for which the laws and constants enable intelligent life to exist (this is the 'fine-tuning' problem). Level III is the multiverse of the many-worlds interpretation of quantum mechanics. Level IV is the multiverse of all possible mathematical structures corresponding to different fundamental laws of physics.

I'll leave you to decide what to make of this.

The many-worlds interpretation and the different varieties of multiverse theory have attracted some high-profile advocates, such as Sean Carroll, Neil deGrasse Tyson, David Deutsch, Brian Greene, Alan Guth, Lawrence Krauss, Andre Linde, Martin Rees, Leonard Susskind, Max Tegmark, Lev Vaidman, and David Wallace. Note once again that this list includes 'neo-Everettians' who do not necessarily interpret the multiverse realistically, but prefer to think about it as a useful conceptual device. Curiously, these tend to be philosophers: it is frequently the scientists who want to be so much more literal. Those raising their voices against this kind of approach—for all kinds of different reasons—include Paul Davies, George Ellis, David Gross, Sabine Hossenfelder, Roger Penrose, Carlo Rovelli, Joe Silk, Paul Steinhardt, Neil Turok, and Peter Woit. I'm inclined to agree with them.[36]

Of course, academic scientists are free to choose what they want to believe and within reason they can publish and say what they like. But in their public pronouncements and publications, the highly speculative and controversial nature of multiverse theories are often overlooked, or simply ignored. The multiverse is cool. Put multiverse in the title or in the headlines of an article and it is more likely to capture attention, get reported in the mainstream media, and invite that all-important click, or purchase.

Many worlds can also be positioned as a rather fashionable rejection of the Copenhagen interpretation, with its advocates (especially Everett and DeWitt) romanticized and portrayed as heroes, 'sticking it to the man', the 'man' in question being Bohr and Heisenberg, and their villainous orthodoxy.[37] This is the picture that Adam Becker paints in his recent popular book *What is Real?* Becker argues that theories 'need to give explanations, unify previously disparate concepts, and bear some relationship with the world around us'.[38] But when all contact with Empirical Reality is lost and all we are left with is the metaphysics, who decides what constitutes 'some relationship'?

In his recent book on quantum mechanics, Lee Smolin calls this tendency 'magical realism',[39] and I personally believe this is very dangerous territory. The temptation to fight dogma with yet more dogma can be hard to resist. When taken together with other speculative theories of foundational physics, the multiverse tempts us away from what many regard as rather old-fashioned notions of the scientific method; they want to wean us off our obsession with empirical evidence and instead just embrace the 'parsimony' that comes with purely metaphysical explanations.

At a time when the authority of science is increasingly questioned by those promoting a firmly anti-scientific agenda, this kind of thing can't be good. As noted Danish historian Helge Kragh concluded:[40]

> But, so it has been argued, intelligent design is hardly less testable than many multiverse theories. To dismiss intelligent design on the ground that it is untestable, and yet to accept the multiverse as an interesting scientific hypothesis, *may come suspiciously close to applying double standards.* As seen from the perspective of some creationists, and also by some non-creationists, their cause has received unintended methodological support from multiverse physics.

Don't get me wrong. I fully understand why those theorists and philosophers who prefer to adopt a realist perspective feel they have no choice but to accept the many-worlds interpretation. But, in the absence of evidence, *personal preferences don't translate into really existing physical things.* For my part I'll happily accept that many worlds were of enormous value *as a way of thinking* about quantum computation. *But thinking about them doesn't make them real.* And, whilst alternative anti-realist interpretations may be less philosophically acceptable to some, it must be admitted that they just don't drag quite so much metaphysical baggage around with them. The formalism itself remains passively neutral and inscrutable. It doesn't care what we think it means.

One last point. Unlike even the more outrageously speculative realist interpretations we've considered thus far, interpretations based on many worlds or the multiverse offer no real clues as to how we might gain any further empirical evidence one way or the other. This is, for me at least, where the multiverse theories really break down. Whatever we think they might be 'explaining' about the nature of quantum reality, we have to admit that there's little or nothing practical to be gained from such explanations. They provide no basis for taking any kind of action according to Proposition #4. Even when predictions are claimed, they're little different from the vague soothsayers' tricks cited by Popper. Unsurprising, really, as this is surely what Einstein was warning us about in 1950, when he explained that the 'passion for understanding' leads to the 'illusion that man is able to comprehend the objective world rationally by pure thought without any empirical foundations—in short, by metaphysics'.[41]

As I explained in Chapter 3, the philosopher James Ladyman suggests that we look to the institutions of science to demarcate between science and non-science, and so defend the integrity of science by excluding claims to objective knowledge based on pure metaphysics. But these institutions haven't so far prevented the publication, in *scientific* journals, of research papers empty of empirical content, filled with speculative theorizing that offers little or no promise of ever making any kind of contact with Empirical Reality. Despite efforts by cosmologist George Ellis and astrophysicist Joe Silk to raise a red flag in 2014 and call on some of these institutions to 'defend the integrity of physics',[42] little has changed. Ladyman seems resigned to this fate: 'Widespread error about fundamentals among experts can and does happen,' he tells me.[43] He believes a correction will come in the long run, when a real scientific breakthrough is made.

Until that happens, we have no choice but to watch in horror as the Ship of Science disappears into the maelstrom. All hands are lost.

EPILOGUE

I've Got a Very Bad Feeling about This

Whatever you make of all this, I think you have to agree that quantum mechanics is an extraordinary theory. It forces us to confront difficult questions about what we think we're doing when we develop a scientific representation of physical reality, and what we expect to get from such a representation. And it forces us to face some simple philosophical truths that were all too easy to ignore in classical mechanics.

I hope I've done enough in this book to explain the nature of our dilemma. We can adopt an anti-realist interpretation in which all our conceptual problems vanish, but which obliges us to accept that we've reached the limit of our ability to access deeper truths about a reality of things-in-themselves. The anti-realist interpretations tell us that there's nothing to see here. Of necessity, they offer no hints as to where we might look to gain some new insights or understanding. They are passive; mute witnesses to the inscrutability of nature.

In contrast, the simpler and more palatable realist interpretations based on local or crypto non-local hidden variables offered plenty of hints and continue to motivate ever more exquisitely subtle experiments. Alas, the evidence is now quite overwhelming and all but the most stubborn of physicists accept that nature denies us this easy way out. If we prefer a realist interpretation, taking the wavefunction and all the conceptual problems this implies at face value, then we're left with what I can only call a choice between unpalatable evils. We can choose de Broglie–Bohm theory, and accept non-local spooky action at a distance. We can choose to add a rather ad hoc spontaneous collapse mechanism, and hope for the best. We can choose to involve consciousness in the mix, conflating one seemingly intractable problem with another. Or we can choose Everett, many worlds, and the multiverse.

In his recent book *Einstein's Unfinished Revolution*, Smolin concludes that quantum mechanics must be incomplete, but that 'realism, in any version, has a price we have to pay to get a new theory that makes complete sense and describes nature correctly and completely'.[1] I leave you to decide which of the realist interpretations we've considered in this book might be worth paying the price (or which is the lesser evil). Smolin remains unconvinced by any and all of them, and has grown weary of arguing the ins and outs of existing approaches. He feels he has no choice but to 'head down into the swamps' in search of new ideas, knowing that 'I will almost certainly fail, but I hope to send back reports to interest and inspire those few others who feel in their bones the cost of our ignorance, of giving up the search too soon.'[2]

For many years, I have on balance preferred Einstein's realism. I have championed Bell's rejection of the Copenhagen orthodoxy (I still do). I trained as an experimentalist, and I'd argue that it's really hard to do experiments of any kind—to intervene, in Hacking's parlance—without a strong belief in the reality of the things you're experimenting with. This is why I think it's fundamentally important to unpack what it means to be a 'realist' based on the four propositions I've set out in Chapters 2 and 3. Few scientists will argue against Propositions #1 (objective reality) and #2 (entity realism). But, though my realist convictions are unshaken, the more I've thought about it, the more I've come to question Proposition #3, the presumption that the base concepts of a theory (such as the quantum wavefunction) necessarily represent real physical states. Over the years I've developed some real doubts.

Like the great philosopher Han Solo, I've got a very bad feeling about this.

I'll leave you with just two reasons for my doubts. One derives simply from the way we routinely apply quantum mechanics. I explained that there is no such thing as the 'right' wavefunction and that, provided we follow the rules, we're perfectly at liberty to express the wavefunction in whatever basis is most suited to our problem. In fact, when we stand back and look hard at what we're doing when we use the quantum formalism, we become aware that, for the most part, we don't make use of the wavefunctions at all. We use mathematical objects such as projection operators, projection amplitudes, and expectation values which *derive* from the wavefunctions and more closely connect with that body of knowledge we call quantum physics. To do this, *we don't need to discover what the wavefunction actually looks like.* We just need to know how one quantum state relates to another, and we get this from our experience of the physics.

Surely, this freedom and flexibility is at odds with any realistic interpretation of the wavefunction. Though I don't necessarily favour Rovelli's relational interpretation, I must admit that his claim that the wavefunction is just a way of 'coding' our experience has started to resonate very deeply.

The second reason is that the results of experiments designed to test Bell's and Leggett's inequalities should cause any realist to stop and think. I've grown very wary of the enormous price we have to pay for *any* realist interpretation not yet ruled out by experiment, and I've grown especially wary of realist interpretations that don't lend themselves to experimental test. Yes, I accept that some of these have been successful in motivating new experimental searches in the spirit of Proposition #4, although on a bad day I confess to some deep suspicions about how these experiments will turn out.

I'm not alone. Although we probably shouldn't read too much into the results, a poll of 33 scientists taken during a conference on the foundations of quantum mechanics held in Traunkirchen, Austria, in 2011, suggested a strong bias towards anti-realist interpretations (see Figure 17).[3] For sure, this is not a statistically significant number of respondents, nor is the sample completely unbiased—the conference was organized by Anton Zeilinger, which may explain the fairly strong preference for information-theoretic interpretations and the absence of votes for de Broglie–Bohm theory and consistent histories. And the leading proponents of the many-worlds interpretation listed in Chapter 10 were notable by their absence. Nevertheless, the results are quite striking.

Perhaps, just as Odysseus himself reasoned, it is better to risk a few crew members by sailing too close to Scylla than risk losing the whole Ship of Science in the whirlpool of metaphysical nonsense that is Charybdis. If we are to learn something new, speculations about our representation of reality *must* connect with the empirical facts of the things-as-they-appear. After all, this is what it means for something to be scientific. I cannot accept an interpretation based on pure metaphysics, no matter how 'parsimonious' this might appear. Just as Dennett argued that accepting Cartesian dualism is giving up, so in my view is accepting many worlds.

There may yet be another way out. I'm pretty confident that quantum mechanics is not the end. Despite its unparalleled success, we know it doesn't incorporate space and time in the

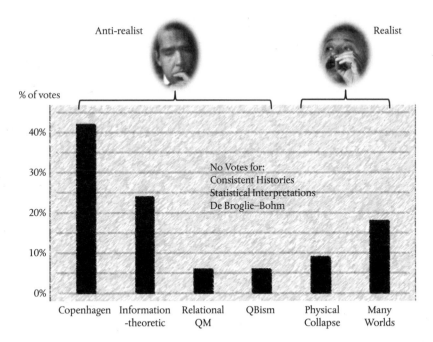

Anti-realist

Realist

% of votes

No Votes for:
Consistent Histories
Statistical Interpretations
De Broglie–Bohm

Copenhagen   Information   Relational   QBism   Physical   Many
             -theoretic   QM   Collapse   Worlds

Figure 17 Results of a multiple-choice questionnaire distributed to 35 attendees of the conference 'Quantum Physics and the Nature of Reality' organized by Anton Zeilinger and held at the International Academy, Traunkirchen, Austria, in July 2011. This graphic shows the results obtained from 33 respondents to the question 'What is your favourite interpretation of quantum mechanics?' I have excluded respondents checking 'Other' (12%) and 'I have no preferred interpretation' (12%). There were no votes for some interpretations I have not considered in this book, such as John Cramer's transactional interpretation.

right way. This is why I don't think closer inspection of a realistically interpreted wavefunction necessarily makes for a good route to an answer, since this tends to drag the baggage of absolute space and time around with it. Unfortunately, we can't look to contemporary efforts to develop a quantum theory of gravity to save us, as these appear to rely too heavily on the existing quantum formalism and don't provide an obvious way to transcend it.

Now it may well be that any theory that transcends quantum mechanics will still be rife with conceptual problems and philosophical conundrums. But it would be nice to discover that, despite appearances to the contrary, there was indeed something more to see here.

APPENDIX

Realist Propositions
and the Axioms of Quantum Mechanics

Realist Propositions (Chapters 2 and 3)

Realist Proposition #1: *The Moon is still there when nobody looks at it (or thinks about it).* There is such a thing as objective reality.

Realist Proposition #2: *If you can spray them, then they are real.* Invisible entities such as photons and electrons really do exist.

Realist Proposition #3: *The base concepts appearing in scientific theories represent the real properties and behaviours of real physical things.* In quantum mechanics, the 'base concept' is the wavefunction.

Realist Proposition #4: *Scientific theories provide insight and understanding, enabling us to do some things that we might otherwise not have considered or thought possible.* This is the 'active' proposition. When deciding whether a theory or interpretation is realist or anti-realist, we ask ourselves what it encourages us to do.

The Axioms of Quantum Mechanics (Chapter 4)

Axiom #1: *The state of a quantum mechanical system is completely defined by its wavefunction.* This is the 'nothing to see here' axiom. The wavefunction has everything you need so don't bother to look for some deeper level of reality that lies beneath it.

Axiom #2: *Observables are represented in quantum theory by a specific class of mathematical operators.* This is the 'right set of keys' axiom.

To get at the observables, such as momentum and energy, we need to unlock the box represented by the wavefunction. Different observables require different keys drawn from the right set.

Axiom #3: *The average value of an observable is given by the expectation value of its corresponding operator.* This is the 'open the box' axiom. It is the recipe we use to combine the operators and the wavefunction to calculate the values of the observables.

Axiom #4: *The probability that a measurement will yield a particular outcome is derived from the square of the corresponding wavefunction.* This is the Born rule, or the 'What might we get?' axiom. Applying the Born rule to a superposition doesn't tell us what we will get from the next measurement.

Axiom #5: *In a closed system with no external influences, the wavefunction evolves in time according to the time-dependent Schrödinger equation.* This is the 'how it gets from here to there' axiom. Note that there's no place here for the kind of discontinuity we associate with the process of measurement. As von Neumann understood, accepting this axiom forces us to adopt a further (but related) axiom in which we assume that a wavefunction representing a superposition of many measurement possibilities collapses to give a single outcome.

ACKNOWLEDGEMENTS

I left academia many years ago, but after spending 13 years studying, teaching, and conducting research in five different universities across the world, I guess certain habits became ingrained and they really do die hard. Wherever I can, I try to get academic specialists to read my stuff and provide comments on it, to reassure me when I've got it right, and to chastise me when I've got it wrong.

Consequently, I'm indebted once again to Carlo Rovelli at Aix-Marseille University for reading the draft manuscript, engaging in extensive correspondence on relational quantum mechanics via email, and for accepting some responsibility for undermining my faith in Einstein's realism. I have also benefited from comments on consistent histories from Robert Griffiths at Carnegie Mellon University; on QBism from Christopher Fuchs at the University of Massachusetts Boston; on the philosophy of science (and my grand metaphor for scientific theorizing) from Massimo Pigliucci at the City College of New York and Michela Massimi at Edinburgh University; on the problem of the preferred basis in the many worlds interpretation from Michael Cuffaro at the University of Western Ontario; on the Everett interpretation from both Harvey Brown at Oxford University and David Wallace at the University of Southern California; and on naturalized metaphysics from James Ladyman at Bristol University. Please understand that my acknowledgement of debt here should *not* lead you to assume that these good folks agree with or are even sympathetic to anything I've written on these subjects in this book. Those errors of confusion and misinterpretation that remain are all my own work.

And, of course, none of this would have been possible without Latha Menon, my long-suffering editor, Jenny Nugee, Lucia Perez, Charles Lauder, and the production team at Oxford University Press, Argentinian artist Eugenia Nobati, who provided the beautifully drawn versions of my metaphor for scientific theorizing (Figure 7) and Wheeler's great smoky dragon (Figure 10), and my son Tim, who provided the pictograms. I will remain eternally grateful for all their efforts.

Jim Baggott
October 2019

LIST OF FIGURE ACKNOWLEDGEMENTS

Figure 12 Reproduced with permission from J. S. Bell, Bertlmann's socks and the nature of reality. *J. Phys. Colloques*, **42** (C2): C2-41-C2-62. Copyright © 1981, EDP Sciences. https://doi.org/10.1051/jphyscol:1981202.

Figure 15 Adapted with permission from C. Philippidis et al. Quantum interference and the quantum potential. *Nuovo Cimento*, **52B**: 15–28. Copyright © 1979, Società Italiana di Fisica.

Figure 16 Adapted with permission from S. Hameroff. Quantum computation in brain microtubules? The Penrose–Hameroff 'Orch OR' model of consciousness. *Philosophical Transactions of the Royal Society of London. Series A: Mathematical, Physical and Engineering Sciences*, **356** (1743). http://doi.org/10.1098/rsta.1998.0254.

Figure 17 The Picture Art Collection / Alamy Stock Photo.

Symbols

🔲 (http://www.onlinewebfonts.com, CC BY 3.0)

🧠 (Premiumvectors / Shutterstock.com)

🔔 (iStock.com / MrPlumo)

ENDNOTES

When referencing the primary scientific literature I've tried as far as possible to provide the original published article *and* the relevant preprint, posted on the online preprint archive managed by Cornell University. The preprints can be accessed free of charge from the arXiv home page—http://arxiv.org/—by typing the article identifier in the search window. Where they are given, direct quotes are typically derived from the preprint.

PREAMBLE

1. I've been developing this metaphor for some time. It features in a short talk I gave in June 2017 on the nature of quantum reality—see https://www.youtube.com/watch?v=VGR68Zl1k8w&. It also features in my book *Quantum Space: Loop Quantum Gravity and the Search for the Structure of Space, Time, and the Universe*, published by Oxford University Press in 2018, and in Jim Baggott, 'The Impossibly Stubborn Question at the Heart of Quantum Mechanics', *Prospect*, 2 August 2018, https://www.prospectmagazine.co.uk/science-and-technology/the-impossibly-stubborn-question-at-the-heart-of-quantum-mechanics
2. Albert Einstein, quoted in Maurice Solovine, *Albert Einstein: Lettres à Maurice Solovine*, Gauthier-Villars, Paris, 1956. This quote is reproduced in Arthur Fine, *The Shaky Game: Einstein, Realism and the Quantum Theory*, 2nd edition (University of Chicago Press, Chicago, 1986), p. 110.

PROLOGUE

1. N. David Mermin, 'A Bolt from the Blue: The E-P-R Paradox', in A. P. French and P. J. Kennedy (eds), *Niels Bohr: A Centenary Volume* (Harvard University Press, Cambridge, MA, 1985), pp. 141–7.

2. In *The Character of Physical Law* (MIT Press, Cambridge, MA, 1967) on p. 129 Richard Feynman famously wrote: 'I think I can safely say that nobody understands quantum mechanics.'

CHAPTER 1: THE COMPLETE GUIDE TO QUANTUM MECHANICS (ABRIDGED)

1. Quoted in Abraham Pais, *Subtle is the Lord: The Science and the Life of Albert Einstein* (Oxford University Press, Oxford, 1982), p. 382.
2. For this demonstration, see https://youtu.be/Iuv6hY6zsdo?t=254
3. Einstein wrote: 'Quantum mechanics is very impressive. But an inner voice tells me that it is not yet the real thing. The theory produces a good deal but hardly brings us closer to the secret of the Old One. I am at all events convinced that *He* does not play dice.' Letter to Max Born, 4 December 1926. Quoted in ibid., p. 443.
4. 'The Unreasonable Effectiveness of Mathematics in the Natural Sciences' was the title of Eugene Wigner's Richard Courant lecture in mathematical sciences delivered at New York University on 11 May 1959. It was published in *Communications on Pure and Applied Mathematics*, **13** (1960), 1–14.
5. Erwin Schrödinger, quoted by Werner Heisenberg in *Physics and Beyond: Memories of a Life in Science* (George Allen & Unwin, London, 1971), p. 75.

CHAPTER 2: JUST WHAT IS THIS THING CALLED 'REALITY', ANYWAY?

1. Steven Weinberg, *Dreams of a Final Theory: The Search for the Fundamental Laws of Nature* (Vintage, London, 1993), p. 133.
2. Lee Smolin and Leonard Susskind, 'Smolin vs. Susskind: The Anthropic Principle', *The Edge*, 18 August 2004: http://www.edge.org/3rd_culture/smolin_susskind04/smolin_susskind.html
3. Lawrence Krauss, presentation to the American Atheists 38th National Convention, 25 March 2012, https://www.youtube.com/watch?v=u9Fi-BqS_Fw. This comment appears around 2:33.

4. And, in any case, as theorist Carlo Rovelli has argued: 'Those who deny the utility of philosophy, are doing philosophy.' See https://blogs.scientificamerican.com/observations/physics-needs-philosophy-philosophy-needs-physics/

5. Larry and Andy Wachowski, *The Matrix: The Shooting Script* (Newmarket Press, New York, 2001), p. 38.

6. Richard E. Cytowic and David M. Eagleman, *Wednesday is Indigo Blue: Discovering the Brain of Synesthesia* (MIT Press, Cambridge, MA, 2009).

7. Thanks to Michela Massimi for spelling this out for me in a personal communication dated 20 March 2019.

8. Philip K. Dick, from the 1978 essay 'How to Build a Universe that Doesn't Fall Apart Two Days Later', included in the anthology *I Hope I Shall Arrive Soon*, edited by Mark Hurst and Paul Williams (Grafton Books, London, 1988). This quote appears on p. 10.

9. Bernard d'Espagnat, *Reality and the Physicist: Knowledge, Duration and the Quantum World* (Cambridge University Press, Cambridge, UK, 1989), p. 115.

10. Karl Popper, quoted in John Horgan, *The End of Science: Facing the Limits of Knowledge in the Twilight of the Scientific Age* (Abacus, London, 1998), p. 35.

11. Werner Heisenberg, *Physics and Philosophy: The Revolution in Modern Science* (Penguin, London, 1989; first published 1958), p. 46.

12. Albert Einstein, quoted in Maurice Solovine, *Albert Einstein: Lettres à Maurice Solovine* (Gauthier-Villars, Paris, 1956). This quote is reproduced in Arthur Fine, *The Shaky Game: Einstein, Realism and the Quantum Theory*, 2nd edition (University of Chicago Press, Chicago, 1986), p. 110.

13. Ian Hacking, *Representing and Intervening: Introductory Topics in the Philosophy of Natural Science* (Cambridge University Press, Cambridge, UK, 1983), p. 23.

14. And I have to confess that I first heard about logical positivism from the infamous Australian 'Bruces' sketch, which appeared in an episode of *Monty Python's Flying Circus* first broadcast in November 1970 (I was 13). This is set in the Philosophy Department of the University of Woolamaloo, in which all the faculty members are called Bruce. 'Now, Bruce teaches classical philosophy, Bruce

teaches Hegelian philosophy, and Bruce here teaches logical positivism and is also in charge of the sheep dip.' From *Monty Python's
Flying Circus: Just the Words*, volume 1 (Mandarin Paperbacks, London,
1990), p. 295.

15. Einstein was forever indebted to Mach for his approach to physics,
but not his aggressively empiricist approach to philosophy. Einstein
once commented that 'Mach was as good at mechanics as he was
wretched at philosophy.' Quoted in Abraham Pais, *Subtle is the Lord:
The Science and the Life of Albert Einstein* (Oxford University Press,
Oxford, 1982). This quote appears on p. 283.

16. For a much more detailed discussion of this aspect of the use of
mathematics in physics, I strongly recommend Giovanni Vignale,
The Beautiful Invisible: Creativity, Imagination, and Theoretical Physics
(Oxford University Press, Oxford, 2011).

17. For an example of a scientist arguing for the acceptance of absolute
spacetime see Brian Greene, *The Fabric of the Cosmos: Space, Time and
the Texture of Reality* (Allen Lane, London, 2004), p. 75 (where he
writes: 'spacetime is a something').

18. Thomas Huxley, 'Biogenesis and Abiogenesis', Presidential Address
to the British Association for the Advancement of Science, 1870,
Collected Essays: Discourses Biological and Geological, volume 8, p. 229.
See: https://mathcs.clarku.edu/huxley/CE8/B-Ab.html. 'Harsh',
'brutal', and 'ugly'—I'm clearly not the first to think that Empirical
Reality is a pretty mean-spirited place.

19. Pierre Duhem, *The Aim and Structure of Physical Theory*, English translation of the second French edition of 1914 by Philip P. Wiener
(Princeton University Press, Princeton, NJ, 1954), p. 145.

CHAPTER 3: SAILING ON THE SEA
OF REPRESENTATION

1. Carl Zimmer, 'In Science, It's Never "Just a Theory"', *New York Times*,
8 April 2016. Available at http://www.nytimes.com/2016/04/09/
science/in-science-its-never-just-a-theory.html?_r=0

2. Bertrand Russell, *The Problems of Philosophy* (Oxford University Press,
Oxford, 1912), p. 35.

3. For an overview of the philosophical programme of naturalized metaphysics, see James Ladyman and Don Ross, with David Spurrett and John Collier, *Every Thing Must Go: Metaphysics Naturalized* (Oxford University Press, Oxford, 2007).

4. Michela Massimi, personal communication, 25 March 2019.

5. Karl Popper, *Conjectures and Refutations: The Growth of Scientific Knowledge* (Routledge & Kegan Paul, London, 1963), p. 49.

6. Lee Smolin, *Time Reborn: From the Crisis in Physics to the Future of the Universe* (Penguin Books, London, 2013), p. 38.

7. Albert Einstein, letter to Paul Ehrenfest, 17 January 1916, quoted in Robert E. Kennedy, *A Student's Guide to Einstein's Major Papers* (Oxford University Press, Oxford, 2012). The quote appears on p. 214.

8. Paul Feyerabend, *Against Method*, 3rd edition (Verso, London, 1993), pp. 52–3.

9. Larry Laudan, 'The Demise of the Demarcation Problem', in R. S. Cohen and L. Laudan (eds), *Physics, Philosophy and Psychoanalysis* (D. Riedel, Dordrecht, 1983), p. 125.

10. Massimo Pigliucci, in Massimo Pigliucci and Maarten Boudry (eds), *The Philosophy of Pseudoscience: Reconsidering the Demarcation Problem* (University of Chicago Press, Chicago, 2013), p. 26.

11. Popper, *Conjectures and Refutations*, p. 346. The italics are mine.

12. Albert Einstein, 'On the Generalised Theory of Gravitation', *Scientific American*, April 1950, p. 13.

13. Don Ross, James Ladyman, and David Spurrett, 'In Defence of Scientism', in Ladyman and Ross, *Every Thing Must Go*, pp. 33–8.

14. See, for example, Lee Smolin, *The Trouble with Physics: The Rise of String Theory, the Fall of a Science, and What Comes Next* (Penguin Books, London, 2008), and Jim Baggott, *Farewell to Reality: How Fairy-tale Physics Betrays the Search for Scientific Truth* (Constable, London, 2013).

15. Gottfried Wilhelm Leibniz, from his correspondence with Samuel Clarke (1715–16), *Collected Writings*, edited by G. H. R. Parkinson (J. M. Dent & Sons, London, 1973), p. 226.

16. Isaac Newton, *Mathematical Principles of Natural Philosophy*, first American edition, translated by Andrew Motte (Daniel Adee, New York, 1845), p. 73.

17. Ernst Mach, *The Science of Mechanics: A Critical and Historical Account of Its Development*, 4th edition, translated by Thomas J. McCormack (Open Court Publishing, Chicago, 1919), p. 194.

18. Albert Einstein, 'Does the Inertia of a Body Depend on its Energy Content?', *Annalen der Physik*, **18** (1905), 639–41. This paper is translated and reproduced in John Stachel (ed.), *Einstein's Miraculous Year: Five Papers That Changed the Face of Physics*, centenary edition (Princeton University Press, Princeton, NJ, 2005). The quote appears on p. 164.

19. See my book *Mass: The Quest to Understand Matter from Greek Atoms to Quantum Fields* (Oxford University Press, Oxford, 2017).

20. See, for example, Richard Boyd, 'On the Current Status of Scientific Realism', *Erkenntnis* **19** (1983), 45–90. Reproduced in Richard Boyd, Philip Gaspar, and J. D. Trout (eds), *The Philosophy of Science* (MIT Press, Cambridge, MA, 1991), see especially p. 195.

21. Hilary Putnam, *Mathematics, Matter and Method* (Cambridge University Press, Cambridge, UK, 1975), p. 73. Quoted in James Ladyman, 'Structural Realism', *Stanford Encyclopedia of Philosophy*, Winter 2016, p. 6.

22. Ian Hacking, *Representing and Intervening: Introductory Topics in the Philosophy of Natural Science* (Cambridge University Press, Cambridge, UK, 1983), p. 31. The italics are mine.

CHAPTER 4: WHEN EINSTEIN CAME DOWN TO BREAKFAST

1. Niels Bohr, quoted by Aage Petersen, 'The Philosophy of Niels Bohr', *Bulletin of the Atomic Scientists*, **19** (1963), 12.

2. A careful analysis of Bohr's philosophical influences and writings suggests that he was closer to the tradition known as *pragmatism* than to positivism. Pragmatism, founded by Charles Sanders Pierce, has many of the characteristics of positivism in that they both roundly reject metaphysics. There are differences, however. We can think of the positivist doctrine as one of 'seeing is believing': what we can know is limited by what we can observe empirically. The pragmatist doctrine admits a more practical (or, indeed, pragmatic) approach: what we can know is limited not by what we

can see, but by what we can *do*. See, for example, Dugald Murdoch, *Niels Bohr's Philosophy of Physics* (Cambridge University Press, Cambridge, UK, 1987).

3. Werner Heisenberg, *The Physical Principles of the Quantum Theory* (University of Chicago Press, Chicago, 1930). Republished in 1949 by Dover Publications, New York. This quote appears in the preface.

4. Max Born and Werner Heisenberg, 'Quantum Mechanics', *Proceedings of the Fifth Solvay Congress, 1928*. English translation from Guido Bacciagaluppi and Antony Valentini, *Quantum Theory at the Crossroads: Reconsidering the 1927 Solvay Conference* (Cambridge University Press, Cambridge, UK, 2009), p. 437.

5. Albert Einstein 'General Discussion', *Proceedings of the Fifth Solvay Congress, 1928*. English translation from Bacciagaluppi and Valentini, *Quantum Theory at the Crossroads*, p. 488.

6. Otto Stern, interview with Res Jost, 2 December 1961. Quoted in Abraham Pais, *Subtle is the Lord: The Science and the Life of Albert Einstein* (Oxford University Press, Oxford, 1982), p. 445.

7. Niels Bohr, in Paul Arthur Schilpp (ed.), 'Discussion with Einstein on Epistemological Problems in Atomic Physics', in *Albert Einstein. Philosopher-scientist*, The Library of Living Philosophers, Volume 1 (Harper & Row, New York, 1959; first published 1949), p. 224.

8. Albert Einstein, quoted by Hendrik Casimir in a letter to Abraham Pais, 31 December 1977. Quoted in Pais, *Subtle is the Lord*, p. 449.

9. Albert Einstein, Boris Podolsky, and Nathan Rosen, 'Can Quantum-Mechanical Description of Physical Reality Be Considered Complete?', *Physical Review*, **47** (1935), 777–80. This paper is reproduced in John Archibald Wheeler and Wojciech Hubert Zurek (eds), *Quantum Theory and Measurement* (Princeton University Press, Princeton, NJ, 1983), pp. 138–41. This quote appears on p. 138.

10. Einstein, Podolsky, and Rosen, 'Can Quantum-Mechanical Description of Physical Reality Be Considered Complete?' Also Wheeler and Zurek (eds), *Quantum Theory and Measurement*, p. 141.

11. Léon Rosenfeld, in Stefan Rozenthal (ed.), *Niels Bohr: His Life and Work as Seen by his Friends and Colleagues* (North-Holland, Amsterdam, 1967), pp. 114–36. Extract reproduced in Wheeler and Zurek (eds), *Quantum Theory and Measurement*, pp. 137 and 142–3. This quote appears on p. 142.

12. Paul Dirac, interview with Niels Bohr, 17 November 1962, *Archive for the History of Quantum Physics.* Quoted in Mara Beller, *Quantum Dialogue* (University of Chicago Press, Chicago, 1999), p. 145.
13. Einstein, Podolsky, and Rosen, 'Can Quantum-Mechanical Description of Physical Reality Be Considered Complete?'. Also in Wheeler and Zurek, (eds), *Quantum Theory and Measurement,* p. 141.
14. Albert Einstein, letter to Erwin Schrödinger, 8 August 1935. Quoted in Arthur Fine, *The Shaky Game: Einstein, Realism and the Quantum Theory,* 2nd edition (University of Chicago Press, Chicago, 1996), p. 78.
15. Erwin Schrödinger, letter to Albert Einstein, 19 August 1935. Quoted ibid., pp. 82–3.
16. Wolfgang Pauli, [review of Dirac, *The Principles of Quantum Mechanics*], *Die Naturwissenschaften,* **19** (1931), 188–9, quoted in Helge Kragh, *Dirac: A Scientific Biography* (Cambridge University Press, Cambridge, UK, 1990), p. 79.

CHAPTER 5: QUANTUM MECHANICS IS COMPLETE SO JUST SHUT UP AND CALCULATE

1. Erwin Schrödinger, letter to Albert Einstein, 7 June 1935. Quoted in Arthur Fine, *The Shaky Game: Einstein, Realism and the Quantum Theory,* 2nd edition (University of Chicago Press, Chicago, 1996), pp. 66–7.
2. Albert Einstein, letter to Erwin Schrödinger, 19 June 1935. Ibid., p. 69.
3. Karl R. Popper, *Quantum Theory and the Schism in Physics* (Unwin Hyman, London, 1982), pp. 99–100.
4. John Bell, quoted by Andrew Whitaker, *John Stewart Bell and Twentieth-Century Physics: Vision and Integrity* (Oxford University Press, Oxford, 2016), p. 57.
5. Lee Smolin, personal communication, 21 June 2017. Quoted in Jim Baggott, *Quantum Space: Loop Quantum Gravity and the Search for the Structure of Space, Time, and the Universe* (Oxford University Press, Oxford, 2018), p. 245. The italics are mine.
6. Carlo Rovelli, 'Relational Quantum Mechanics', *International Journal of Theoretical Physics,* **35** (1996), 1637; arXiv: quant-ph/9609002v2, 24 February 1997, p. 1.
7. Rovelli, ibid., p. 3.

8. Matteo Smerlak and Carlo Rovelli, 'Relational EPR', *Foundations of Physics*, **37** (2007), 427–45; arXiv:quant-ph/0604064v3, 4 March 2007, p. 3.
9. Carlo Rovelli, personal communication, 14 October 2018.
10. Smerlak and Rovelli, 'Relational EPR', arXiv:quant-ph/0604064v3, p. 5.
11. Rovelli, 'Relational Quantum Mechanics', arXiv: quant-ph/9609002v2, p. 4.
12. Anton Zeilinger, 'A Foundational Principle for Quantum Mechanics', *Foundations of Physics*, **29** (1999), 633.
13. See Jeffrey Bub, 'Quantum Mechanics Is about Quantum Information', *Foundations of Physics*, **35** (2005), 541–60. See also arXiv:quant-ph/0408020v2, 12 August 2004.
14. Smerlak and Rovelli, 'Relational EPR', arXiv:quant-ph/0604064v3, p. 5.
15. Ibid., p. 4.
16. Niels Bohr, quoted by Aage Petersen, 'The Philosophy of Niels Bohr', *Bulletin of the Atomic Scientists*, **19** (1963), 12. The italics are mine.
17. A. J. Ayer, in A. J. Ayer (ed.), *Logical Positivism*, Library of Philosophical Movements (Free Press of Glencoe, 1959), p. 11. The italics are mine.
18. Ludwig Wittgenstein, *Tractatus Logico-Philosophicus*, translated by C. K. Ogden (Kegan Paul, Trench, Trubner, London, 1922), p. 90.
19. N. David Mermin, 'Could Feynman Have Said This?', *Physics Today*, May 2004, pp. 10–11.

CHAPTER 6: QUANTUM MECHANICS IS COMPLETE BUT WE NEED TO REINTERPRET WHAT IT SAYS

1. Lucien Hardy, 'Quantum Theory from Five Reasonable Axioms', arXiv:quant-ph/0101012v4, 25 September 2001.
2. Giulio Chiribella, quoted by Philip Ball in 'Quantum Theory Rebuilt from Simple Physical Principles', *Quanta*, 30 August 2017.
3. Karl R. Popper, *Quantum Theory and the Schism in Physics* (Unwin Hyman, London, 1982), p. 72.
4. Robert Griffiths, personal communication, 5 November 2018.
5. Robert B. Griffiths, *Consistent Quantum Theory* (Cambridge University Press, Cambridge, UK, 2002), p. 214.

6. Robert B. Griffiths, 'The Consistent Histories Approach to Quantum Mechanics', *Stanford Encyclopedia of Philosophy*, Spring 2017, p. 3.

7. Fay Dowker and Adrian Kent, 'On the Consistent Histories Approach to Quantum Mechanics', *Journal of Statistical Physics*, **82** (1996), 1575–1646. See also arXiv:gr-qc/9412067v2, 25 January 1996.

8. Griffiths, 'The Consistent Histories Approach to Quantum Mechanics', p. 46.

9. Werner Heisenberg, *Physics and Philosophy: The Revolution in Modern Science* (Penguin, London, 1989; first published 1958), p. 46. The italics are mine.

10. Carlton M. Caves, Christopher A. Fuchs, and Rüdiger Schack, 'Quantum Probabilities as Bayesian Probabilities', *Physical Review A*, **65** (2002), 022305. See also arXiv:quant-ph/0106133v2, 14 November 2001.

11. Richard Healey, 'Quantum-Bayesian and Pragmatist Views of Quantum Theory', *Stanford Encyclopedia of Philosophy*, Spring 2017, p. 9.

12. N. David Mermin, 'Annotated Interview with a QBist in the Making', arXiv:quant-ph/1301.6551.v1, 28 January 2013. However, Mermin interprets 'QBism' a little differently, preferring to acknowledge Bruno de Finetti, a pioneer of subjective probability, rather than Bayes. The 'B' then stands for 'Bruno'.

13. Schack has argued that Hardy's 'five reasonable axioms' can be reduced to four, by reinterpreting the first in terms of Bayesian probabilities and by modifying part of the proof. See Rüdiger Schack, 'Quantum Theory from Four of Hardy's Axioms', *Foundations of Physics*, **33** (2003), 1461–8. See also arXiv:quant-ph/0210017v1, 2 October 2002.

14. Christopher A. Fuchs, N. David Mermin, and Rüdiger Schack, 'An Introduction to QBism with an Application to the Locality of Quantum Mechanics', *American Journal of Physics*, **82** (2014), 749–54. See also arXiv:quant-ph/1311.5253v1, 20 November 2013.

15. Mermin talks about CBism, the classical analogue of QBism. See N. David Mermin, 'Making Better Sense of Quantum Mechanics', arXiv:quant-ph/1809.01639v1, 5 September 2018.

16. Chrisopher A. Fuchs, 'On Participatory Realism', arXiv:quant-ph/1601.04360v3, 28 June 2016, p. 11.

CHAPTER 7: QUANTUM MECHANICS IS INCOMPLETE SO WE NEED TO ADD SOME THINGS

1. John Bell, *Journal de Physique* Colloque C2, Supplement 3, **42** (1981), 41–61. Reproduced in J. S. Bell, *Speakable and Unspeakable in Quantum Mechanics* (Cambridge University Press, Cambridge, UK, 1987), pp. 139–58. This quote appears on p. 142.

2. Darrin W. Belousek, 'Einstein's 1927 Unpublished Hidden-Variable Theory: Its Background, Context and Significance', *Studies in History and Philosophy of Science Part B: Studies in History and Philosophy of Modern Physics*, **27** (1996), 437–61. Peter Holland takes a closer look at Einstein's reasons for rejecting this approach in 'What's Wrong with Einstein's 1927 Hidden-Variable Interpretation of Quantum Mechanics', *Foundations of Physics*, **35** (2005), 177–96; arXiv:quant-ph/0401017v1, 5 January 2004.

3. 'It should be noted that we need not go any further into the mechanism of the "hidden parameters", since we now know that the established results of quantum mechanics can never be re-derived with their help.' John von Neumann, *Mathematical Foundations of Quantum Mechanics* (Princeton University Press, Princeton, NJ, 1955), p. 324.

4. According to Basil Hiley, one of Bohm's long-term collaborators, Bohm said of his meeting with Einstein: 'After I finished [*Quantum Theory*] I felt strongly that there was something seriously wrong. Quantum theory had no place in it for an adequate notion of an individual actuality. My discussions with Einstein clarified and reinforced my opinion and encouraged me to look again.' Quoted by Basil Hiley, personal communication to the author, 1 June 2009.

5. David Bohm, *Quantum Theory* (Prentice-Hall, Englewood Cliffs, NJ, 1951), p. 623.

6. D. Bohm and Y. Aharonov, 'Discussion of Experimental Proof for the Paradox of Einstein, Rosen, and Podolsky', *Physical Review*, **108** (1957), 1070.

7. John Bell, in P. C. W. Davies and J. R. Brown (eds), *The Ghost in the Atom* (Cambridge University Press, Cambridge, UK, 1986), p. 57.

8. John Bell, 'Bertlmann's Socks and the Nature of Reality', *Journal de Physique* Colloque C2, Supplement 3, **42** (1981), 41–61. Reproduced in

Bell, *Speakable and Unspeakable in Quantum Mechanics: Collected Papers on Quantum Philosophy* (Cambridge University Press, Cambridge, UK, 1987), pp. 139–58. This quote appears on p. 139.

9. This pictorial representation is based on Bernard d'Espagnat, 'The Quantum Theory and Reality', *Scientific American*, **241** (1979), 158–81. See p. 162.

10. John Bell, 'Locality in Quantum Mechanics: Reply to Critics', *Epistemological Letters*, November 1975, pp. 2–6. This paper is reproduced in Bell, *Speakable and Unspeakable in Quantum Mechanics*, pp. 63–6. This quote appears on p. 65.

11. Simon Kochen and E. P. Specker, 'The Problem of Hidden Variables in Quantum Mechanics', *Journal of Mathematics and Mechanics*, **17** (1967), 59–87.

12. John S. Bell, 'On the Problem of Hidden Variables in Quantum Theory', *Reviews of Modern Physics*, **38** (1966), 447–52. This paper is reproduced in Bell, *Speakable and Unspeakable in Quantum Mechanics*, pp. 1–13.

13. Calcite is a naturally birefringent form of calcium carbonate. It has a crystal structure which has different refractive indices along two distinct crystal planes. One offers an axis of maximum transmission for vertically polarized light and the other offers an axis of maximum transmission for horizontally polarized light. The vertical and horizontal components of either left- or right-circularly polarized light are therefore physically separated by passage through the crystal, and their intensities can be measured separately. With careful machining, a calcite crystal can transmit virtually all of the light incident on it.

14. Alain Aspect, Philippe Grangier, and Gérard Roger, 'Experimental Tests of Realistic Local Theories via Bell's Theorem', *Physical Review Letters*, **47** (1981), 460–3. Alain Aspect, Philippe Grangier, and Gérard Roger, 'Experimental Realization of Einstein–Podolsky–Rosen–Bohm *Gedankenexperiment*: A New Violation of Bell's Inequalities', *Physical Review Letters*, **49** (1982), 91–4.

15. Alain Aspect, Jean Dalibard, and Gérard Roger, 'Experimental Test of Bell's Inequalities Using Time-Varying Analyzers', *Physical Review Letters*, **49** (1982), 1804–7.

16. W. Tittel, J. Brendel, N. Gisin, and H. Zbinden, 'Long-Distance Bell-Type Tests Using Energy-Time Entangled Photons', *Physical Review A*, **59** (1999), 4150–63.

17. Thomas Scheidl, Rupert Ursin, Johannes Kofler, Sven Ramelow, Xiao-Song Ma, Thomas Herbst, Lothar Ratschbacher, Alessandro Fedrizzi, Nathan K. Langford, Thomas Jennewein, and Anton Zeilinger, 'Violation of Local Realism with Freedom of Choice', *Proceedings of the National Academy of Sciences*, **107** (2010), 19708–13.

18. Dominik Rauch, Johannes Handsteiner, Armin Hochrainer, Jason Gallicchio, Andrew S. Friedman, Calvin Leung, Bo Liu, Lukas Bulla, Sebastian Ecker, Fabian Steinlechner, Rupert Ursin, Beili Hu, David Leon, Chris Benn, Adriano Ghedina, Massimo Cecconi, Alan H. Guth, David I. Kaiser, Thomas Scheidl, and Anton Zeilinger, 'Cosmic Bell Test Using Random Measurement Settings from High-Redshift Quasars', *Physical Review Letters*, **121** (2018), 080403.

19. Jian-Wei Pan, Dik Bouwmeester, Matthew Daniell, Harald Weinfurter, and Anton Zeilinger, 'Experimental Test of Quantum Nonlocality in Three-Photon Greenburger-Horne-Zeilinger Entanglement', *Nature*, **403** (2000), 515–19.

20. A. J. Leggett, 'Nonlocal Hidden-Variable Theories and Quantum Mechanics: An Incompatibility Theorem', *Foundations of Physics*, **33** (2003), 1469–93. This quote appears on pp. 1474–5.

21. Simon Gröblacher, Tomasz Paterek, Rainer Kaltenbaek, Caslav Brukner, Marek Zukowski, Markus Aspelmeyer, and Anton Zeilinger, 'An Experimental Test of Non-local Realism', *Nature*, **446** (2007), 871–5. In case you were wondering, Bell's inequality is violated in these experiments, too.

22. Matthew F. Pusey, Jonathan Barrett, and Terry Rudolph, 'On the Reality of the Quantum State', *Nature Physics*, **8** (2012), 475–8.

23. For an excellent overview, see Matthew Saul Leifer, 'Is the Quantum State Real? An Extended Overview of ψ–ontology Theorems', *Quanta*, **3** (2014), 67–155.

CHAPTER 8: QUANTUM MECHANICS IS INCOMPLETE SO WE NEED TO ADD SOME OTHER THINGS

1. See, for example, Guiseppe Pucci, Daniel M. Harris, Luiz M. Faria, and John W. M. Bush, 'Walking Droplets Interacting with Single and Double Slits', *Journal of Fluid Mechanics*, **835** (2018), 1136–56.

2. See Natalie Wolchover, 'Famous Experiment Dooms Alternative to Quantum Weirdness', *Quanta Magazine*, 11 October 2018: https://www.quantamagazine.org/famous-experiment-dooms-pilot-wave-alternative-to-quantum-weirdness-20181011/.

3. Peter R. Holland, *The Quantum Theory of Motion: An Account of the de Broglie-Bohm Causal Interpretation of Quantum Mechanics* (Cambridge University Press, Cambridge, UK, 1993), p. 475.

4. Ibid., p. 462.

5. Albert Einstein, letter to Max Born, 12 May 1952. Quoted in John S. Bell, *Proceedings of the Symposium on Frontier Problems in High Energy Physics*, Pisa, June 1976, pp. 33–45. This paper is reproduced in J. S. Bell, *Speakable and Unspeakable in Quantum Mechanics: Collected Papers on Quantum Philosophy* (Cambridge University Press, Cambridge, UK, 1987), pp. 81–92. The quote appears on p. 91.

6. See James T. Cushing, *Quantum Mechanics: Historical Contingency and the Copenhagen Hegemony* (University of Chicago Press, Chicago, 1994).

7. J. S. Bell, 'On the Impossible Pilot Wave', *Foundations of Physics*, **12** (1982), 989–99. This paper is reproduced in Bell, *Speakable and Unspeakable in Quantum Mechanics*, pp. 159–68.

8. J. S. Bell, 'Against Measurement', *Physics World*, **3** (1990), 33.

9. These estimates are taken from Roland Omnès, *The Interpretation of Quantum Mechanics* (Princeton University Press, Princeton, NJ, 1994). The original calculations were reported in E. Joos and H. D. Zeh, *Zeitschrift für Physik*, **B59** (1985), 223–43.

10. For some examples, see Serge Haroche, 'Entanglement, Decoherence and the Quantum/Classical Boundary', *Physics Today*, July 1998, 36–42.

11. Frédéric Bouchard, Jérémie Harris, Harjaspreet Mand, Nicolas Bent, Enrico Santamato, Robert W. Boyd, and Ebrahim Karimi, 'Observation of Quantum Recoherence of Photons by Spatial Propagation', *Nature Scientific Reports* (2015), 5:15330.

12. Markus Arndt, Olaf Nairz, Julian Voss-Andreae, Claudia Keller, Gerbrand van der Zouw, and Anton Zeilinger, 'Wave-particle Duality of C_{60} molecules', *Nature*, **401** (1999), 680–2; Markus Arndt, Olaf Nairz, J. Petschinka, and Anton Zeilinger, 'High Contrast Interference with C_{60} and C_{70}', *Comptes Rendus de l'Académie des*

Sciences—Series IV—Physics, **2** (2001), 581–5; and Stefan Gerlich, Sandra Eibenberger, Mathias Tomandl, Stefan Nimmrichter, Klaus Hornberger, Paul J. Fagan, Jens Tüxen, Marcel Mayor, and Markus Arndt, 'Quantum Interference of Large Organic Molecules', *Nature Communications* (2011), 2:263. The analogies and puns are endless. The ratio of the diameter of a C_{60} molecule and the spacing of the silicon nitride grating used to observe the diffraction pattern is comparable to the ratio of the diameter of a conventional soccer ball and the width of a goal (according to FIFA standards), giving a potentially whole new meaning to the term 'bend it like Beckham'.

13. See, for example, Jonathan R. Friedman, Vijay Patel, W. Chen, S. K. Tolpygo, and J. E. Lukens, 'Quantum Superposition of Distinct Macroscopic States', *Nature*, **406** (2000), 43–6, and Caspar H. van der Wal, A. C. J. ter Haar, F. K. Wilhelm, R. N. Schouten, C. J. P. M. Harmans, T. P. Orlando, Seth Lloyd, and J .E. Moonij, 'Quantum Superposition of Macroscopic-Persistent States', *Science*, **290** (2000), 773–7.

14. Bell, 'Against Measurement', 33.

15. Roger Penrose, *The Large, the Small and the Human Mind*, Canto edition (Cambridge University Press, Cambridge, UK, 2000), p. 82. Note that the collapse of the wavefunction is sometimes referred to as the 'reduction' of the wavefunction.

16. For example, in the second edition of David J. Griffiths' popular textbook *Introduction to Quantum Mechanics*, published by Cambridge University Press in 2017, decoherence is mentioned just once, in a footnote.

17. See G. C. Ghirardi, A. Rimini, and T. Weber, 'Unified Dynamics for Microscopic and Macroscopic Systems', *Physical Review D*, **34** (1986), 470–91; and P. Pearle, 'Combining Stochastic Dynamical State-Vector Reduction with Spontaneous Localization', *Physical Review A*, **39** (1989), 2277–89. Although they're not entirely equivalent, the 'state vector' referred to in the title of this paper can be taken to be similar to the wavefunction.

18. J. S. Bell, 'Are There Quantum Jumps?', in C. W. Kilmister (ed.), *Schrödinger: Centenary Celebration of a Polymath* (Cambridge University Press, Cambridge, UK, 1987), pp. 41–52. This article is reproduced in

Bell, *Speakable and Unspeakable in Quantum Mechanics*, pp. 201–12. This quote appears on p. 204.

19. Giancarlo Ghirardi, 'Collapse Theories', *Stanford Encylopedia of Philosophy*, Substantive Revision, February 2016, p. 51.

20. John Wheeler, with Kenneth Ford, *Geons, Black Holes and Quantum Foam: A Life in Physics* (W.W. Norton, New York, 1998), p. 235.

21. L. Diósi, 'A Universal Master Equation for the Gravitational Violation of Quantum Mechanics', *Physical Letters A*, **120** (1987), 377–81; L. Diósi, 'Models for Universal Reduction of Macroscopic Quantum Fluctuations', *Physical Review A*, **40** (1989), 1165–74; and Roger Penrose, 'On Gravity's Role in Quantum State Reduction', *General Relativity and Gravitation*, **28** (1996), 581–600.

22. Roger Penrose, *The Emperor's New Mind: Concerning Computers, Minds and the Laws of Physics* (Vintage, London, 1990), p. 475.

23. See Jim Baggott, *Quantum Space: Loop Quantum Gravity and the Search for the Structure of Space, Time, and the Universe* (Oxford University Press, Oxford, 2018), pp. 259–62.

24. Roger Penrose, *Fashion, Faith and Fantasy in the New Physics of the Universe* (Princeton University Press, Princeton, NJ, 2016), p. 215.

25. Rainer Kaltenbaek, Gerald Hechenblaikner, Nikolai Kiesel, Oriol Romero-Isart, Keith C. Schwab, Ulrich Johann, and Markus Aspelmeyer, 'Macroscopic Quantum Resonators (MAQRO)', *Experimental Astronomy*, **34** (2012), 123–64, see also arXiv:quant-ph/1201.4756v2, 19 March 2012.

26. Rainer Kaltenbaek, et al., 'Macroscopic Quantum Resonators (MAQRO): 2015 Update', *EPJ Quantum Technology*, **3** (2016), 5.

CHAPTER 9: QUANTUM MECHANICS IS INCOMPLETE BECAUSE WE NEED TO INCLUDE MY MIND (OR SHOULD THAT BE YOUR MIND?)

1. John von Neumann, *Mathematical Foundations of Quantum Mechanics* (Princeton University Press, Princeton, NJ, 1955), p. 420.

2. Ibid., p. 421.

3. L. Szilard, 'On Entropy Reduction in a Thermodynamic System by Interference by Intelligent Beings', *Zeitschrift fur Physik*, **53** (1929), 840–56. NASA Technical Translation F-16723.

4. Max Jammer, *The Philosophy of Quantum Mechanics* (Wiley, New York, 1974), p. 480. The italics are mine.

5. Eugene Wigner was another Hungarian compatriot. Together with Szilard and Edward Teller, Wigner was part of the 'Hungarian conspiracy' that influenced Einstein to write a letter to US President Franklin D. Roosevelt on 2 August 1939 warning of 'extremely powerful bombs of a new type'. See Jim Baggott, *Atomic: The First War of Physics and the Secret History of the Atom Bomb 1939–49* (Icon Books, London, 2009), pp. 18–19.

6. Eugene Wigner, 'Remarks on the Mind-Body Question', in I. J. Good (ed.), *The Scientist Speculates: An Anthology of Partly-Baked Ideas* (Heinemann, London, 1961), pp. 284–302. This is reproduced in John Archibald Wheeler and Wojciech Hubert Zurek (eds), *Quantum Theory and Measurement* (Princeton University Press, Princeton, NJ, 1983), pp. 168–81. These quotes appear on pp. 176–8.

7. John Archibald Wheeler, 'Law without Law', in Wheeler and Zurek (eds), *Quantum Theory and Measurement*, pp. 182–213. This quote appears on p. 184.

8. Ibid., p. 185.

9. John Archibald Wheeler, 'Genesis and Observership', in Robert E. Butts and Jaakko Hintikka (eds), *Foundational Problems in the Special Sciences* (D. Reidel, Dordrecht, Holland, 1977), p. 28. Wheeler was referring to Robert Dicke, who highlighted the 'fine tuning' in physical laws and constants that seems to be necessary in order for life to be possible in the Universe, and Brandon Carter, who developed the anthropic principle in 1974 as a direct challenge to the Copernican principle.

10. John Wheeler, with Kenneth Ford, *Geons, Black Holes and Quantum Foam: A Life in Physics* (W.W. Norton, New York, 1998), p. 338.

11. John D. Barrow and Frank Tipler, *The Anthropic Cosmological Principle* (Oxford University Press, Oxford, 1986), p. 22. Italics in the original.

12. David J. Chalmers, 'Facing up to the problem of consciousness', *Journal of Consciousness Studies*, **2** (1995), 200–19.

13. Neuroscientists Stephen Macknik and Susana Martinez-Conde explore the neuroscience of magic in their entertaining book *Sleights of Mind*, published by Profile Books, London, 2011.

14. Gilbert Ryle, *The Concept of Mind* (Hutchinson, London, 1949).

15. Daniel Dennett, *Consciousness Explained* (Penguin, London, 1991).
16. Not all neuroscientists agree. For a lucid and witty rebuttal of the reductionist approach, I recommend Raymond Tallis, *Aping Mankind: Neuromania, Darwinitis, and the Misrepresentation of Humanity* (Routledge Classics, London, 2016). For alternative 'top-down' arguments, see George Ellis, *How Can Physics Underlie the Mind: Top-Down Causation in the Human Context* (Springer-Verlag, Berlin, 2016).
17. Cambridge Declaration on Consciousness, 7 July 2012.
18. Roger Penrose, *The Emperor's New Mind: Concerning Computers, Minds, and the Laws of Physics* (Vintage, London, 1990), p. 540.
19. Ibid., p. 517.
20. Recent computer simulations of the molecular dynamics of a guanosine diphosphate (GDP) bound tubulin made use of a structure consisting of 13,432 atoms, not counting the 150,510 water molecules which surrounds and 'solvates' it. See Yeshitila Gebremichael, Jhih-Wei Chu, and Gregory A. Voth, 'Intrinsic Bending and Structural Rearrangement of Tubulin Dimer: Molecular Dynamics Simulations and Coarse-Grained Analysis', *Biophysical Journal*, **95** (2008), 2487–99.
21. Max Tegmark, 'The Importance of Quantum Decoherence in Brain Processes', *Physical Review E*, **61** (2000), 4194–206. See also arXiv:quant-ph/9907009v2, 10 November 1999.
22. Satyajit Sahu, Subrata Ghosh, Kazuto Hirata, Daisuke Fujita, and Anirban Bandyopadhyay, 'Multi-level Memory-Switching Properties of a Single Brain Microtubule', *Applied Physics Letters*, **102** (2013), 123701.
23. Stuart Hameroff and Roger Penrose, 'Consciousness in the Universe: A Review of the "Orch-OR" Theory', *Physics of Life Reviews*, **11** (2014), 70.
24. Chalmers, 'Facing up to the problem of consciousness', pp. 200–19. The italics are mine.

CHAPTER 10: QUANTUM MECHANICS IS
INCOMPLETE BECAUSE . . . OKAY, I GIVE UP

1. Albert Einstein, 'Approximative Integration of the Field Equations of Gravitation', *Preussische Akademie der Wissenschaften (Berlin) Sitzungsberichte*, 1916, 688–96. Quoted in Gennady E. Gorelik and

Viktor Ya. Frenkel, *Matvei Petrovich Bronstein and Soviet Theoretical Physics in the Thirties* (Birkhäuser Verlag, Basel, 1994). The quote appears on p. 86.

2. Hugh Everett III, '"Relative State" Formulation of Quantum Mechanics', *Reviews of Modern Physics*, **29** (1957), 454–62. This is reproduced in John Archibald Wheeler and Wojciech Hubert Zurek (eds), *Quantum Theory and Measurement* (Princeton University Press, Princeton, NJ), 1983, pp. 315–23. This quote appears on p. 316. Italics in the original. I should point out that Everett left academia and joined the Pentagon's Weapons System Evaluation Group in June 1956, and although this paper doesn't carry Wheeler's name it represents a compromise that Everett was never entirely happy with.

3. Hugh Everett III, 'The Theory of the Universal Wave Function', Princeton University PhD Thesis. This is reproduced in B. S. DeWitt and N. Graham (eds), *The Many Worlds Interpretation of Quantum Mechanics* (Pergamon Press, Oxford, 1975). The italics are mine.

4. Everett, 'The Theory of the Universal Wave Function', footnote on p. 68. Italics in the original.

5. Stefano Osnaghi, Fábio Freitas, and Olival Freire Jr, 'The Origin of the Everettian Heresy', *Studies in the History and Philosophy of Modern Physics*, **40** (2009), 111.

6. Everett, 'The Theory of the Universal Wave Function', p. 9.

7. Nancy G. Everett, letter to Frank J. Tipler, 10 October 1983. Quoted by Eugene Shikhovtsev, 'Biographical Sketch of Hugh Everett, III', available in an online version maintained by Max Tegmark: http://space.mit.edu/home/tegmark/everett/everett.html

8. Bryce S. DeWitt, 'Quantum Mechanics and Reality', *Physics Today*, **23** (1970), 30. This is reproduced in DeWitt and Graham (eds), *The Many Worlds Interpretation of Quantum Mechanics*.

9. John Wheeler, in P. C. W. Davies and J. R. Brown (eds), *The Ghost in the Atom: A Discussion of the Mysteries of Quantum Physics* (Cambridge University Press, Cambridge, UK, 1986), p. 60.

10. J. A. Wheeler, letter to Paul Benioff, 7 July 1977. Quoted by Shikhovtsev, 'Biographical Sketch of Hugh Everett, III'.

11. DeWitt, 'Quantum Mechanics and Reality', p. 33.

12. This is an observation frequently attributed to John Wheeler, but I picked it up from the interview with David Deutsch which appears in Davies and Brown, *The Ghost in the Atom*, p. 84.

13. H. D. Zeh, 'The Problem of Conscious Observation in Quantum Mechanical Description', *Foundations of Physics Letters*, **13** (2000), 221–33. See also arXiv:quant-ph/9908084v3, 5 June 2000. Zeh explains that this paper is an update of a paper that was first informally circulated through the *Epistemological Letters* of the Ferdinand-Gonseth Association in Biel, Switzerland, in 1981.

14. Michael Lockwood, *Mind, Brain and the Quantum. The Compound 'I'* (Blackwell, Oxford, 1990).

15. David Z. Albert, *Quantum Mechanics and Experience* (Harvard University Press, Cambridge, MA, 1992).

16. J. S. Bell, 'Quantum Mechanics for Cosmologists', in C. Isham, R. Penrose, and D. Sciama (eds), *Quantum Gravity 2* (Clarendon Press, Oxford, 1981). This paper is reproduced in J. S. Bell, *Speakable and Unspeakable in Quantum Mechanics* (Cambridge University Press, Cambridge, UK, 1987), pp. 117–38. This quote appears on p. 118.

17. Murray Gell-Mann, *The Quark and the Jaguar: Adventures in the Simple and the Complex* (Little, Brown, London, 1994), p. 138.

18. Adrian Kent, 'One World versus Many: The Inadequacy of Everettian Accounts of Evolution, Probability, and Scientific Confirmation', in Simon Saunders, Jonathan Barrett, Adrian Kent, and David Wallace (eds), *Many Worlds? Everett, Quantum Theory, & Reality* (Oxford University Press, Oxford, 2010), p. 310.

19. David Deutsch, *The Fabric of Reality* (Allen Lane, London, 1997), p. 45.

20. Ibid., p. 53.

21. Ibid., p. 216.

22. Max Tegmark, 'The Interpretation of Quantum Mechanics: Many Worlds or Many Words?', *Fortschritte der Physik*, **46** (1998), 855–62. See also arXiv:quant-ph/9709032v1, 15 September 1997. This quote appears on p. 5.

23. David Wallace, *The Emergent Multiverse: Quantum Theory According to the Everett Interpretation* (Oxford University Press, Oxford, 2012), p. 158.

24. Ibid., p. 371.

25. Lev Vaidman, 'Review: David Wallace, *The Emergent Multiverse*', *British Journal for the Philosophy of Science*, **66** (2014), 465–8. See also Lev

Vaidman, 'Many-Worlds Interpretation of Quantum Mechanics', *Stanford Encyclopedia of Philosophy*, substantive revision 17 January 2014.

26. David Wallace, 'Decoherence and Ontology (or: How I Learned to Stop Worrying and Love FAPP)', in Simon Saunders, Jonathan Barrett, Adrian Kent, and David Wallace (eds), *Many Worlds? Everett, Quantum Theory, & Reality* (Oxford University Press, Oxford, 2010), p. 62.

27. R. Raussendorf and H. J. Briegel, 'A One-Way Quantum Computer', *Physics Review Letters*, **86** (2001), 5188–91.

28. P. Walther, K. J. Resch, T. Rudolph, E. Schenck, H. Weinfurter, V. Vedral, M. Aspelmeyer, and A. Zeilinger, 'Experimental One-Way Quantum Computing', *Nature*, **434** (2005), 169–76. See also arXiv:quant-ph/0503126, 14 March 2005.

29. Michael Cuffaro, 'Many Worlds, the Cluster State Quantum Computer, and the Problem of the Preferred Basis', *Studies in History and Philosophy of Modern Physics*, **43** (2012), 35–42. See also arXiv:physics. hist-ph/1110.2514v2, 10 January 2012. This quote appears on p. 17.

30. Michael Cuffaro, personal communication, 19 June 2019.

31. David Wallace, personal communication, 25 June 2019.

32. David Wallace, personal communication, 27 June 2019.

33. David Deutsch, interview with John Horgan, 'The Infinite Optimism of Physicist David Deutsch', 17 January 2018, https://blogs.scientificamerican.com/cross-check/the-infinite-optimism-of-physicist-david-deutsch/

34. Martin Rees, 'What are the Limits of Human Understanding?', *Prospect Magazine*, 13 November 2018. https://www.prospectmagazine.co.uk/magazine/martin-rees-what-are-the-limits-of-human-understanding

35. Max Tegmark, *Our Mathematical Universe: My Quest for the Ultimate Nature of Reality* (Penguin Books, London, 2015); see particularly Chapter 8.

36. See for example Jim Baggott, *Farewell to Reality: How Fairy-tale Physics Betrays the Search for Scientific Truth* (Constable, London, 2013), especially Chapter 9; and Sabine Hossenfelder, *Lost in Math: How Beauty Leads Physics Astray* (Basic Books, New York, 2018).

37. See Christopher A. Fuchs, 'Copenhagen Interpretation Delenda Est?', arXiv:quant-ph/1809.05147v2, 11 November 2018.

38. Adam Becker, *What is Real? The Unfinished Quest for the Meaning of Quantum Physics* (John Murray, London, 2018), p. 264.

39. Lee Smolin, *Einstein's Unfinished Revolution: The Search for What Lies beyond the Quantum* (Allen Lane, London, 2019), p. xxiii.

40. Helge Kragh, *Higher Speculations: Grand Theories and Failed Revolutions in Physics and Cosmology* (Oxford University Press, Oxford, 2011), p. 285. The italics are mine.

41. Albert Einstein, 'On the Generalized Theory of Gravitation', *Scientific American*, **182** (April 1950), p. 13. Einstein continues: 'I believe that every true theorist is a kind of tamed metaphysicist, no matter how pure a "positivist" he may fancy himself. The metaphysicist believes that the logically simple is also the real. The tamed metaphysicist believes that not all that is logically simple is embodied in experienced reality, but that the totality of all sensory experience can be "comprehended" on the basis of a conceptual system built on premises of great simplicity.'

42. George Ellis and Joe Silk, *Nature*, 516 (2014), 321–3.

43. James Ladyman, personal communication, 29 March 2019.

EPILOGUE: I'VE GOT A VERY BAD FEELING ABOUT THIS

1. Lee Smolin, *Einstein's Unfinished Revolution: The Search for What Lies beyond the Quantum* (Allen Lane, London, 2019), p. 180.

2. Ibid., p. 277.

3. Maximilian Schlosshauer, Johannes Koer, and Anton Zeilinger, 'A Snapshot of Foundational Attitudes toward Quantum Mechanics', arXiv:quant-ph/1301.1069v1, 6 January 2013. Before you ask, no, I did not attend. But I was extremely gratified to be invited to talk at a follow-up conference on the same subject, held in Traunkirchen in November 2013, following publication of my book *Farewell to Reality*.

BIBLIOGRAPHY

Here's an interesting question. In a popular science book, what is meant to be the *purpose* of a bibliography? To provide some reassurance to the reader that the book is well researched and thorough, and based on the publications of recognized authorities on the subject? Or it is supposed to demonstrate the cleverness of the author—look at what I had to read in order to bring you this book?

In making my decisions on what to include here, I've chosen to imagine an inquisitive reader who might be encountering the conceptual and philosophical problems of quantum mechanics for the first time, and who might be sufficiently motivated by what they read in this book to want to find out more. The result is the kind of bibliography that would have been useful to me back in 1987 when, in exasperation, I wanted to know why somebody hadn't told me about all this before.

POPULAR SCIENCE

These are books that anyone with an interest but no formal education in quantum physics should be able to pick up and enjoy. I've biased this list towards more recent publications, but there are some golden oldies that are just too good to be excluded.

ACZEL, AMIR D., *Entanglement: The Greatest Mystery in Physics* (Wiley, London, 2003).

ALBERT, DAVID Z., *Quantum Mechanics and Experience* (Harvard University Press, Cambridge, MA, 1992).

ANANTHASWAMY, ANIL, *Through Two Doors at Once: The Elegant Experiment that Captures the Enigma of Our Quantum Reality* (Dutton, New York, 2018).

BALL, PHILIP, *Beyond Weird: Why Everything You Thought You Knew About Quantum Physics is Different* (The Bodley Head, London, 2018).

BAGGOTT, JIM, *The Quantum Story: A History in 40 Moments* (Oxford University Press, Oxford, 2011).

BAGGOTT, JIM, *Farewell to Reality: How Fairy-tale Physics Betrays the Search for Scientific Truth* (Constable, London, 2013).

BAGGOTT, JIM, *Quantum Space: Loop Quantum Gravity and the Search for the Structure of Space, Time, and the Universe* (Oxford University Press, Oxford, 2018).

BARROW, JOHN D., and TIPLER, FRANK J., *The Anthropic Cosmological Principle* (Oxford University Press, Oxford, 1986).

BECKER, ADAM, *What is Real: The Unfinished Quest for the Meaning of Quantum Physics* (John Murray, London, 2018).

CARROLL, SEAN, *Something Deeply Hidden: Quantum Worlds and the Emergence of Spacetime* (Oneworld, London, 2019).

DAVIES, P. C. W., and BROWN, J. R. (EDS), *The Ghost in the Atom: A Discussion of the Mysteries of Quantum Physics* (Cambridge University Press, Cambridge, UK, 1986).

DEUTSCH, DAVID, *The Fabric of Reality* (Penguin, London, 1997).

FEYNMAN, RICHARD, *The Character of Physical Law* (MIT Press, Cambridge, MA, 1967).

FEYNMAN, RICHARD P., *QED: The Strange Theory of Light and Matter* (Penguin, London, 1985).

GAMOW, GEORGE, *Thirty Years That Shook Physics* (Dover, New York, 1966).

GELL-MANN, MURRAY, *The Quark and the Jaguar* (Little, Brown, London, 1994).

GREENE, BRIAN, *The Elegant Universe: Superstrings, Hidden Dimensions and the Quest for the Ultimate Theory* (Vintage Books, London, 2000).

GREENE, BRIAN, *The Fabric of the Cosmos: Space, Time and the Texture of Reality* (Allen Lane, London, 2004).

GREENE, BRIAN, *The Hidden Reality: Parallel Universes and the Deep Laws of the Cosmos* (Allen Lane, London, 2011).

GRIBBIN, JOHN, *Schrödinger's Kittens* (Penguin, London, 1995).

KUMAR, MANJIT, *Quantum: Einstein, Bohr and the Great Debate About the Nature of Reality* (Icon Books, London, 2008).

LINDLEY, DAVID, *Where Does the Weirdness Go?* (Basic Books, New York, 1996).

ORZEL, CHAD, *How to Teach Quantum Physics to Your Dog* (Oneworld, London, 2010).

PENROSE, ROGER, *The Emperor's New Mind: Concerning Computers, Minds and the Laws of Physics* (Vintage, London, 1990).

PENROSE, ROGER, *Shadows of the Mind: A Search for the Missing Science of Consciousness* (Vintage, London, 1995).

PENROSE, ROGER, *The Road to Reality: A Complete Guide to the Laws of the Universe* (Vintage, London, 2005).

PENROSE, ROGER, *Fashion, Faith and Fantasy in the New Physics of the Universe* (Princeton University Press, Princeton, NJ, 2016).

RAYMER, MICHAEL G., *Quantum Physics: What Everyone Needs to Know®* (Oxford University Press, Oxford, 2017).

RAE, ALASTAIR, *Quantum Physics: Illusion or Reality?* (Cambridge University Press, Cambridge, UK, 1986).

REES, MARTIN, *Just Six Numbers: The Deep Forces that Shape the Universe* (Phoenix, London, 2000).

ROVELLI, CARLO, *Seven Brief Lessons on Physics* (Allen Lane, London, 2015).

ROVELLI, CARLO, *Reality is Not What it Seems: The Journey to Quantum Gravity* (Allen Lane, London, 2016).

ROVELLI, CARLO, *The Order of Time* (Allen Lane, London, 2018).

SACHS, MENDEL, *Einstein versus Bohr: The Continuing Controversies in Physics* (Open Court, La Salle, IL, 1988).

SMOLIN, LEE, *Three Roads to Quantum Gravity: A New Understanding of Space, Time and the Universe* (Phoenix, London, 2001).

SMOLIN, LEE, *The Trouble with Physics: The Rise of String Theory, the Fall of a Science, and What Comes Next* (Penguin Books, London, 2008).

SMOLIN, LEE, *Time Reborn: From the Crisis in Physics to the Future of the Universe* (Penguin Books, London, 2014).

SMOLIN, LEE, *Einstein's Unfinished Revolution: The Search for What Lies beyond the Quantum* (Allen Lane, London, 2019).

SUSSKIND, LEONARD, *The Cosmic Landscape: String Theory and the Illusion of Intelligent Design* (Little, Brown, New York, 2006).

TEGMARK, MAX, *Our Mathematical Universe: My Quest for the Ultimate Nature of Reality* (Penguin Books, London, 2015).

VON BAEYER, HANS CHRISTIAN, *QBism: The Future of Quantum Physics* (Harvard University Press, Cambridge, MA, 2016).

WEINBERG, STEVEN, *Dreams of a Final Theory: The Search for the Fundamental Laws of Nature* (Vintage, London, 1993).

WILCZEK, FRANK, *The Lightness of Being: Big Questions, Real Answers* (Allen Lane, London, 2009).

SCIENTIFIC BIOGRAPHIES

I've always believed that one of the best ways to try to understand how quantum mechanics developed to its present form and why it continues to baffle new generations of students and consumers of popular science is to study the biographies of those responsible for it. You quickly realize that all the clever arguments—one way or the other—were already being made in the late 1920s, by scientists who really did know what they were talking about. But—watch out—many of these biographies are technical and require some background in quantum mechanics.

BERNSTEIN, JEREMY, *Quantum Profiles* (Princeton University Press, Princeton, NJ, 1991). Contains biographical sketches of John Bell and John Wheeler.

BIRD, KAI, and SHERWIN, MARTIN J., *American Prometheus: The Triumph and Tragedy of J. Robert Oppenheimer* (Atlantic Books, London, 2008).

CASSIDY, DAVID C., *Uncertainty: The Life and Science of Werner Heisenberg* (W. H. Freeman, New York, 1992).

ENZ, CHARLES P., *No Time to be Brief: A Scientific Biography of Wolfgang Pauli* (Oxford University Press, Oxford, 2002).

FARMELO, GRAHAM, *The Strangest Man: The Hidden Life of Paul Dirac, Quantum Genius* (Faber & Faber, London, 2009).

FEYNMAN, RICHARD P., *'Surely You're Joking, Mr. Feynman!'* (Unwin, London, 1985).

GLEICK, JAMES, *Genius: Richard Feynman and Modern Physics* (Little, Brown, London, 1992).

GOODCHILD, PETER, *J. Robert Oppenheimer* (BBC, London, 1980).

HEILBRON, J. L., *The Dilemmas of an Upright Man: Max Planck and the Fortunes of German Science* (Harvard University Press, Cambridge, MA, 1996).

HEISENBERG, WERNER, *Physics and Beyond: Memories of a Life in Science* (George Allen & Unwin, London, 1971).

HOFFMANN, BANESH, *Albert Einstein* (Paladin, St. Albans, 1975).

HORGAN, JOHN, 'Last Words of a Quantum Heretic', *Scientific American*, February 1993, p. 38. In part a biographical sketch of David Bohm.

ISAACSON, WALTER, *Einstein: His Life and Universe* (Simon & Schuster, New York, 2007).

KLEIN, MARTIN J., *Paul Ehrenfest: The Making of a Theoretical Physicist*, Vol. 1, 3rd edn (North-Holland, Amsterdam, 1985).

KRAGH, HELGE S., *Dirac: A Scientific Biography* (Cambridge University Press, Cambridge, UK, 1990).

MACRAE, NORMAN, *John von Neumann: The Scientific Genius Who Pioneered the Modern Computer, Game Theory, Nuclear Deterrence, and Much More* (Pantheon Books, New York, 1992).

MEHRA, JAGDISH, *The Beat of a Different Drum: The Life and Science of Richard Feynman* (Oxford University Press, Oxford, 1994).

MONK, RAY, *Inside the Centre: The Life of J. Robert Oppenheimer* (Vintage, London, 2013).

MOORE, WALTER, *Schrödinger: Life and Thought* (Cambridge University Press, Cambridge, UK, 1989).

NASAR, SYLVIA, *A Beautiful Mind* (Faber and Faber, London, 1998). A biography of the mathematician John Nash, containing a biographical sketch of John von Neumann.

PAIS, ABRAHAM, *Subtle is the Lord: The Science and the Life of Albert Einstein* (Oxford University Press, Oxford, 1982).

PAIS, ABRAHAM, *Niels Bohr's Times, in Physics, Philosophy, and Polity* (Oxford University Press, Oxford, 1991).

PAIS, ABRAHAM, *J. Robert Oppenheimer: A Life* (Oxford University Press, Oxford, 2006).

PEAT, F. DAVID, *Infinite Potential: The Life and Times of David Bohm* (Addison-Wesley, Reading, MA, 1997).

POUNDSTONE, WILLIAM, *Prisoner's Dilemma: John von Neumann, Game Theory and the Puzzle of the Bomb* (Random House, New York, 1992).

WHEELER, JOHN ARCHIBALD, with Ford, Kenneth, *Geons, Black Holes and Quantum Foam: A Life in Physics* (W. W. Norton, New York, 1998).

WHITAKER, ANDREW, *John Stewart Bell and Twentieth-Century Physics: Vision and Integrity* (Oxford University Press, Oxford, 2016).

ADVANCED TEXTS

Specifically for those who are considering studying, are studying, or have studied quantum mechanics at university. I've included here some relevant books on the philosophy of science and the philosophy of quantum mechanics.

AYER, A. J., *Language, Truth, and Logic* (Penguin, London, 1936).

BAGGOTT, JIM, *The Quantum Cookbook: Mathematical Recipes for the Foundations of Quantum Mechanics* (Oxford University Press, Oxford, 2020).

BELLER, MARA, *Quantum Dialogue: The Making of a Revolution* (University of Chicago Press, Chicago, 1999).

BOHM, DAVID, *Quantum Theory* (Prentice-Hall, Englewood Cliffs, NJ, 1951).

BOHM, DAVID, *Causality and Chance in Modern Physics* (Routledge, London, 1957).

BOHM, D., and HILEY, B. J., *The Undivided Universe* (Routledge, London, 1993).

CARTWRIGHT, NANCY, *How the Laws of Physics Lie* (Oxford University Press, 1983).

CUSHING, JAMES T., *Quantum Mechanics: Historical Contingency and the Copenhagen Hegemony* (University of Chicago Press, Chicago, 1994).

CUSHING, JAMES T., *Philosophical Concepts in Physics* (Cambridge University Press, Cambridge, UK, 1998).

DE BROGLIE, LOUIS, *Matter and Light: The New Physics*, translated by W. H. Johnston (W. W. Norton, New York, 1939).

D'ESPAGNAT, BERNARD, *The Conceptual Foundations of Quantum Mechanics*, 2nd edn (Addison-Wesley, New York, 1989).

D'ESPAGNAT, BERNARD, *Reality and the Physicist: Knowledge, Duration and the Quantum World* (Cambridge University Press, Cambridge, UK, 1989).

DIRAC, P. A. M., *The Principles of Quantum Mechanics*, 4th edn (Clarendon Press, Oxford, 1958).

DUHEM, PIERRE, *The Aim and Structure of Physical Theory*, translated by Philip P. Wiener (Princeton University Press, Princeton, NJ, 1954).

FEYERABEND, PAUL, *Farewell to Reason* (Verso, London, 1987).

FEYERABEND, PAUL, *Against Method*, 3rd edn (Verso, London, 1993).

FEYNMAN, RICHARD P., LEIGHTON, ROBERT B., and SANDS, MATTHEW, *The Feynman lectures on Physics*, Vol. III (Addison-Wesley, Reading, MA, 1965).

FINE, ARTHUR, *The Shaky Game: Einstein, Realism and the Quantum Theory*, 2nd edn (University of Chicago Press, Chicago, 1986).

GARDNER, SEBASTIAN, *Kant and the Critique of Pure Reason* (Routledge, London, 1999).

GILLIES, DONALD, *Philosophy of Science in the Twentieth Century* (Blackwell, Oxford, 1993).

GRIFFITHS, DAVID J., *Introduction to Quantum Mechanics*, 2nd edn (Cambridge University Press, Cambridge, UK, 2017).

HACKING, IAN, *Representing and Intervening* (Cambridge University Press, Cambridge, UK, 1983).

HEISENBERG, WERNER, *The Physical Principles of the Quantum Theory* (Dover, New York, 1930).

HEISENBERG, WERNER, *Physics and Philosophy: The Revolution in Modern Science* (Penguin, London, 1989; first published 1958).

HOLLAND, PETER R., *The Quantum Theory of Motion: An Account of the De Broglie-Bohm Causal Interpretation of Quantum Mechanics* (Cambridge University Press, Cambridge, UK, 1993).

ISHAM, CHRIS J., *Lectures on Quantum Theory* (Imperial College Press, London, 1995).

JAMMER, MAX, *The Philosophy of Quantum Mechanics* (Wiley, New York, 1974).

KRAGH, HELGE, *Quantum Generations: A History of Physics in the Twentieth Century* (Princeton University Press, Princeton, NJ, 1999).

KRAGH, HELGE, *Higher Speculations: Grand Theories and Failed Revolutions in Physics and Cosmology* (Oxford University Press, Oxford, 2011).

KRAGH, HELGE, *Niels Bohr and the Quantum Atom: The Bohr Model of Atomic Structure 1913–1925* (Oxford University Press, Oxford, 2012).

KUHN, THOMAS S., *The Structure of Scientific Revolutions*, 2nd edn (University of Chicago Press, Chicago, 1970).

KUHN, THOMAS S., *Black-body Theory and the Quantum Discontinuity 1894–1912* (University of Chicago Press, Chicago, 1978).

LADYMAN, JAMES, and ROSS, DON, with Spurrett, David and Collier, John, *Every Thing Must Go: Metaphysics Naturalized* (Oxford University Press, Oxford, 2007).

LAKATOS, IMRE, and MUSGRAVE, ALAN (EDS), *Criticism and the Growth of Knowledge* (Cambridge University Press, Cambridge, UK, 1970).

LOCKWOOD, MICHAEL, *Mind, Brain and the Quantum. The Compound 'I'* (Blackwell, Oxford, 1990).

MEHRA, JAGDISH, *Einstein, Physics and Reality* (World Scientific, London, 1999).

MURDOCH, DUGALD, *Niels Bohr's Philosophy of Physics* (Cambridge University Press, Cambridge, UK, 1987).

OMNÈS, ROLAND, *The Interpretation of Quantum Mechanics* (Princeton University Press, Princeton, NJ, 1994).

OMNÈS, ROLAND, *Understanding Quantum Mechanics* (Princeton University Press, Princeton, NJ, 1999).

OMNÈS, ROLAND, *Quantum Philosophy* (Princeton University Press, Princeton, NJ, 1999).

POPPER, KARL R., *The Logic of Scientific Discovery* (Hutchinson, London, 1959).

POPPER, KARL, *Conjectures and Refutations: The Growth of Scientific Knowledge* (Routledge & Kegan Paul, London, 1963).

POPPER, KARL R., *Quantum Theory and the Schism in Physics* (Unwin Hyman, London, 1982).

PRESTON, JOHN, *Kuhn's The Structure of Scientific Revolutions: A Reader's Guide* (Continuum, London, 2008).

PSILLOS, STATHIS, *Scientific Realism: How Science Tracks Truth* (Routledge, London, 1999).

RAE, ALASTAIR I. M., *Quantum Mechanics*, 2nd edn (Adam Hilger, Bristol, 1986).

RUSSELL, BERTRAND, *The Problems of Philosophy* (Oxford University Press, Oxford, 1967).

RYLE, GILBERT, *The Concept of Mind* (Penguin, London, 1963).

SAUNDERS, SIMON, BARRETT, JONATHAN, KENT, ADRIAN, and WALLACE, DAVID (EDS), *Many Worlds? Everett, Quantum Theory, & Reality* (Oxford University Press, Oxford, 2010).

SCHACHT, RICHARD, *Classical Modern Philosophers: Descartes to Kant* (Routledge & Kegan Paul, London, 1984).

VAN FRAASSEN, BAS C., *The Scientific Image* (Oxford University Press, Oxford, 1980).

VON NEUMANN, JOHN, *Mathematical Foundations of Quantum Mechanics* (Princeton University Press, Princeton, NJ, 1955).

WALLACE, DAVID, *The Emergent Multiverse: Quantum Theory According to the Everett Interpretation* (Oxford University Press, Oxford, 2012).

ANTHOLOGIES AND COLLECTIONS

Where necessary, I've put references to the 'primary' scientific literature—meaning original research papers and review articles—in the Endnotes. But some of the most important papers in the history and philosophy of quantum mechanics have been collected together and published in book form, which makes them nicely accessible for interested readers who don't have ready access to a university library. Here is a selection of some of the best.

BELL, J. S., *Speakable and Unspeakable in Quantum Mechanics* (Cambridge University Press, Cambridge, UK, 1987).

BORN, MAX, *Physics in My Generation*, 2nd edn (Springer, New York, 1969).

DEWITT, B. S., and GRAHAM, N. (EDS), *The Many Worlds Interpretation of Quantum Mechanics* (Pergamon, Oxford, 1975).

FRENCH, A. P., and KENNEDY, P. J. (EDS), *Niels Bohr: A Centenary Volume* (Harvard University Press, Cambridge, MA, 1985). This includes the chapter by N. David Mermin, 'A Bolt from the Blue: The E-P-R Paradox', which got me started back in 1987.

HAWKING, STEPHEN (ED.), *The Dreams that Stuff Is Made Of: The Most Astounding Papers on Quantum Physics and How They Shook the World* (Running Press, Philadelphia, 2011).

HEISENBERG, WERNER, *Encounters with Einstein* (Princeton University Press, Princeton, NJ, 1983).

HILEY, B. J., and PEAT, F. D. (EDS), *Quantum Implications: Essays in Honour of David Bohm* (Routledge & Kegan Paul, London, 1987).

KILMISTER, C. W. (ED.), *Schrödinger: Centenary Celebration of a Polymath* (Cambridge University Press, Cambridge, UK, 1987).

OPPENHEIMER, J. ROBERT, *Atom and Void* (Princeton University Press, Princeton, NJ, 1989).

SCHILPP, P. A. (ED.), *Albert Einstein: Philosopher-Scientist*. The Library of Living Philosophers (Open Court, La Salle, IL, 1949). This includes a chapter by Bohr on their famous debate about the representation of quantum reality.

STACHEL, JOHN (ED.), *Einstein's Miraculous Year: Five Papers that Changed the Face of Physics* (Princeton University Press, Princeton, NJ, 2005).

VAN DER WAERDEN, B. L., *Sources of Quantum Mechanics* (Dover, New York, 1968).

WHEELER, JOHN ARCHIBALD, *At Home in the Universe* (AIP Press, New York, 1994).

WHEELER, JOHN ARCHIBALD, and ZUREK, WOJCIECH HUBERT (EDS), *Quantum Theory and Measurement* (Princeton University Press, Princeton, NJ, 1983).

INDEX

n after a page number indicates a footnote or an endnote. f after a page number indicates a figure.

acceleration 74
Adams, Douglas 141
Adams, John 64
Aharonov, Yakir 155, 156, 167
Albert, David 230
angular momentum 155
anthropic principle 206–7, 277
anti-realist interpretations 249, 252
arc-seconds 65n
Aspect, Alain 167, 171
atomic spectra 10f
atoms 7–8
axioms 99–102, 127–30, 255–6
Ayer, Alfred J. 124

Bacon, Francis 57
Bandyopadhyay, Anirban 219
Barrett, Jonathan 173
Barrow, John 207
base concepts 71
basis states 113
Bayes, Thomas 142
Bayesian probability theory 142–3
Becker, Adam 245
Bell, John 106, 151, 157–67, 182, 183, 189, 230–1
Bell's inequality 157, 164–7, 170, 171
Bertlmann, Reinhard 158
Big Bang model 108
black-body radiation 8
Bohm, David 154–5, 156, 167, 176, 177, 271
Bohr, Niels
    atomic theory 9
    interpretation of quantum theory 28, 81, 82–4, 85, 86–7,
        88–91, 94–5, 98, 106, 138, 185, 202, 226
    philosophical views 124, 266
Bohr–Einstein debate 84, 85–96
Boltzmann, Ludwig 96
Born, Max 22, 84, 240
Born rule 101, 129–31, 138
bosons 167
Briegel, Hans 242
Bub, Jeffrey 122
Bush, John 178

C_{60} molecules 275
calcite 272
calcium atoms 168
Carnap, Rudolph 47
Carter, Brandon 277
cathode ray guns 15
Caves, Carlton 144
CERN 45, 108, 158
Chalmers, David 209, 219
chemistry 26
Chiribella, Giulio 129
Christiansen, Christian 10f
classical mechanics 2, 15n, 17
cloud chambers 12
cluster states 241–2
collapse of the wavefunction 24,
    181–9, 200, 202, 221–2
complementarity 83–4, 106, 107
complex conjugates 22n
Compton, Arthur 60
consciousness 207–13
consistent histories
    interpretation 135–8
constructive empiricism 80

Cooper, Leon 229
Copenhagen interpretation 84
    challenges/opposition 95, 98, 131,
        148, 226, 227
    influence/legacy 98, 102, 105–7,
        138, 154, 155, 175, 176, 182
    limitations 86, 106–7, 140, 203,
        228n
Copernican principle 144, 277
Crane, Louis 109
crypto non-local hidden variable
    theories 173–4, 249
Cuffaro, Michael 242

D-Wave machines 237n
d'Espagnat, Bernard 41
Davy, Humphrey 51
de Broglie, Louis 2n, 10, 14f, 31, 175
de-Broglie–Bohm interpretation
    176–82, 250
decoherence 139–40, 179, 185–90,
    192, 230, 231, 241
decoherence times 186–7, 219
decoherent histories 139
decoherent histories interpretation
    140, 231
Debye, Pieter 60
delocalization 12, 21
demarcation problem 62–7
Dennett, Daniel 211
Descartes, René 209–10
Deutsch, David 233–4, 243
Dewdney, Chris 177
DeWitt, Bryce 227, 228, 229, 233
Dick, Philip K. 40
Dicke, Robert 277
diffraction 12, 188
Diósi, Lajos 195
Diósi–Penrose theory 195–6, 214
Dirac, Paul 94, 99
Dirac's principles 99–102
Dowker, Fay 140, 141
Duhem, Pierre 52, 64
Duhem–Quine thesis 64

Einstein, Albert
    general theory of relativity 50, 65,
        222
    influence of Mach 264
    interpretation of quantum
        theory 24, 44, 84, 85–91, 93, 94,
        95–7, 105, 152–4, 155, 181,
        262, 271
    letter to Franklin D. Roosevelt
        277
    light-quantum hypothesis 9,
        59–60
    Nobel Prize 11
    philosophical views 43, 44n, 69,
        109, 282
Ellis, George 247
energy 72
energy–time uncertainty
    relation 30, 88
ensembles 183–4
entangled particles 91, 118
entropy 184–5, 191
EPR (Einstein–Podolsky–Rosen)
    experiment 91–6, 105, 118, 155,
        156, 157, 179
Euclid's axioms 102
European Space Agency (ESA) 197
Everett, Hugh 223–7, 233, 279
expectation values 18

factorization 236n
falsifiability criterion 62, 66
Faraday, Michael 51
fermions 156
Feyerabend, Paul 66
Feynman, Richard 4, 125, 195
'five reasonable axioms' 128, 270
'force particles' 167
Fuchs, Christopher 144, 145
functional magnetic resonance
    imaging (fMRI) 211–12

Galle, Johann 64
Gell-Mann, Murray 139, 140, 231–2

general theory of relativity 50, 65, 90, 108, 222
Ghirardi, Giancarlo 193, 194
'ghost in the machine' 211
Graham, Neill 228, 229
grand unified theory 154, 195
gravitational waves 108
gravity 108–9
'greater than or equal to' sign 29n
Griffiths, Robert 135, 136, 137, 140, 141
Grover, Lov 242n
'guiding field' (*Führungsfeld*) 153

Hacking, Ian 46, 79
Hameroff, Stuart 214, 215, 216, 219
Hamilton, William Rowan 2
Hamiltonian mechanics 2
hard problem (consciousness) 209, 212
Hardy, Lucien 128–9
Hartle, James 139, 140
Heisenberg, Werner 28, 42, 81, 84, 106, 143–4, 202
Heisenberg's uncertainty principle 28–30, 90
Hempel, Carl 57n
hidden variables theories 154–5
Higgs, Peter 68
Higgs boson 68, 108
Higgs field 75
Hilbert, David 99
Hilbert space 134n, 200n
Hilbert's problems 99
Hiley, Basil 177, 271
Hitch-hiker's Guide to the Galaxy (Douglas Adams) 141
Holland, Peter 180
Huxley, Thomas 51
hypothetico-deductive method 60

imaginary numbers 17n
information 110–22, 190–2

see also relational/information-theoretic interpretations
information theory 191
intelligent design 63
intensity (waves) 21
interference 12, 13f, 19–21, 137, 138, 188
intrinsic behaviour 21
irreversibility 185

Kant, Immanuel 39–40, 44n
Kent, Adrian 140, 141, 232
Kochen, Simon 165
Kragh, Helge 246
Krauss, Lawrence 34
Kuhn, Thomas 80n

'la Comédie Française' 15
Ladyman, James 69, 247
Lagrange, Joseph-Louis 2
Landauer, Rolf 191
Laplace, Pierre Simon 142
Laudan, Larry 66
Le Verrier, Urbain 64, 65
Leggett, Anthony 172–3
Leggett's inequality 173
light-quantum hypothesis 9, 10, 12, 60
linear momentum 15, 16, 17
LISA Pathfinder mission 197–8
localization accuracy 193
Lockwood, Michael 230
Loewer, Barry 230
logical positivism 47, 62, 263
loop quantum gravity 108

Mach, Ernst 47, 50, 264
macroscopic quantum resonators (MAQRO) 197, 198, 218
Mandelstam, Leonid 30
many-minds theory 230
many-worlds interpretation 228–47, 252
mass 75

Massimi, Michela 61
materialism 211
Matrix movie 35
'matter waves' 22
Maxwell, James Clerk 96
Maxwell's Demon 203
measurement operator 23
mechanics 2
Mercator projection 134
Mercury (planet) 65
Mermin, N. David 3, 125, 145
metaphysical preconceptions 50
metaphysics 34
microtubules 215
Miller, Kenneth R. 55–6
mind–body dualism 210–11
Möbius band 156n
modulus square (wavefunctions) 22
Molière 44
Monty Python's Flying Circus 263–4
Monty Python's Life of Brian 35
Moore's law 236n
multiverse theories 233–40, 243–5,
    246–7

naïve realism 43
neo-Everettians 244
neural correlates 211
Neurath, Otto 47
neuroscience 212
New York Times 94
Newton, Isaac 60, 74
Newton's first law of motion 180
Newton's second law of motion 72
Newton's third law of motion 205
Newtonian precession 65
no-go theorems 165, 173

observables 17, 18
observers 109–10
Omnès, Roland 140, 190
operators (mathematics) 18
orchestrated objective reduction
    (Orch-OR) 214, 216–18, 219

Pais, Abraham 228n
participatory anthropic principle
    206–7
Pauli, Wolfgang 84, 94, 99, 106
PBR (Pusey–Barrett–Rudolph)
    theorem 174, 176
Pearle, Philip 193
Penrose, Roger 190, 195–6, 197,
    214–15, 216, 219
Philippidis, Chris 177
photoelectric effect 11
'photon box' experiment 88–90
photons 9, 167
Pierce, Charles Sanders 266
Pigliucci, Massimo 67
pilot wave theories 175–6, 182, 230
Planck, Max 8–9
Planck's constant 9, 16, 17
Planck–Einstein relation 8, 11
Plato 38
Podolsky, Boris 91, 93
Poincaré, Henri 79
polarizing analysers 169
Popper, Karl
    definition of physics 41
    interpretation of quantum
        theory 105–6, 131–2, 133
    philosophy of science 58–9, 60,
        62, 67
positron emission tomography
    (PET) 211, 212
posterior probabilities 143
pragmatism 266–7
principle of induction 57, 62
prior probabilities 142
projection operators 134
prokaryotic cells 215n
Pusey, Mathew 173
Putnam, Hilary 77

quanta 8
quantum Bayesianism (QBism)
    145–9
quantum computing 235–7, 241–3

quantum jumps 9
quantum measurement 183, 189
quantum numbers 9
quantum potential 176–81
quantum probabilities 22
quantum probability theory 24
quantum theory of gravity 109, 195,
   222
qubits 235, 241, 242n
Quine, Willard Van Orman 64

Raussendorf, Robert 242
raven paradox 57n
realist interpretations 249–50
realist propositions 44, 46, 71, 79, 255
recoherence 188
red roses 208, 211
Rees, Martin 244
relational/information-theoretic
   interpretations 109–11, 115,
   122–5, 148, 191, 251
Rimini, Alberto 193
Roosevelt, Franklin D. 277
Rosen, Nathan 91, 93
Rosenfeld, Léon 222
Ross, Don 69
Rovelli, Carlo 107–9, 110, 113–14, 116,
   119, 122, 196n
Rudolph, Terry 173
Russell, Bertrand 58
Ryle, Gilbert 211

Schack, Rüdiger 144, 145, 270
Schlick, Moritz 47
Schrödinger, Erwin
  development of quantum
    theory 16–17, 26, 31, 84–5
  interpretation of quantum
    theory 22, 28, 44, 81, 95, 96,
    97–8, 105
Schrödinger's cat 97–8, 116, 123, 212,
   229
Schrödinger's wave equation 16, 27f,
   130, 176

scientific method 56
scientific realism 43
Searle, John 145
second law of thermodynamics
   184
Shannon, Claude 191
Shannon entropy 191
'shifty split' 106, 136, 190
'shut up and calculate' slogan 124
Silk, Joe 247
Sisyphus (Greek mythology) 72
Smolin, Lee 64, 108–9, 196n, 245,
   250
social brain hypothesis 213
Solo, Han 251
Solvay conferences 84, 87, 175
spacetime 50, 222–3
special theory of relativity 181
Specker, Ernst 165
spin (elementary particles) 155–6
'spooky action at a distance' 85,
   94, 180
Spurrett, David 69
Standard Model 108
state vectors 275
statistical interpretation 152–4
Stern, Otto 86–7
Stern–Gerlach apparatus 159n
structural realism 78
superconducting quantum interference
   devices (SQUIDs) 188
superposition 26, 29
superstring theory 243
Susskind, Leonard 34
synapses (brain) 215
synesthesia 37
Szilard, Leo 202, 277

Tamm, Igor 30
Tegmark, Max 219, 238, 239, 244
Teller, Edward 277
thought experiments
   (gedankenexperiments) 87
Tipler, Frank 207

total wavefunction 91
Traunkirchen conference (2011) 252, 253f
tubulin subunits 215, 216–18, 278
two-slit experiment
    see interference

uncertainty principle 28–30, 90
Uranus 63, 64

Vaidman, Lev 240
van der Waals forces 216
van Fraassen, Bas 80
van Vechten, Deborah 229
verifiability criterion 62
Vienna Circle 47, 124
von Neumann, John 23, 99, 154, 184, 199–202, 203, 221

Wallace, David 240, 241, 243
Watt, Richard 215
wave equation
    see Schrödinger's wave equation

wave mechanics 16–19
wave–particle duality 16
wavefunctions 16, 17, 22, 26–7, 251
wavepackets 29
Weber, Tullio 193
Weinberg, Steven 34
Wheeler, John
    collaboration with Hugh
        Everett 223, 226, 228, 233
    description of general relativity
        theory 195
    development of participatory
        anthropic principle 206–7
    interpretation of quantum
        theory 123–4, 185
Wigner, Eugene 204–5, 277
Wigner's friend 205
Wittgenstein, Ludwig 124
Worrall, John 78

Zeh, Dieter 185, 230, 231
Zeilinger, Anton 120, 121, 252, 253f